Working Donkeys in 4th-3rd Millennium BC Mesopotamia

Working Donkeys in 4th-3rd Millennium BC Mesopotamia: Insights from Modern Development Studies is a reassessment of the role and impact of working-animal adoption in antiquity, focusing on 4th-3rd millennium BC Mesopotamia but applicable to other periods and regions.

This book is driven by a novel interdisciplinary process of analogy with modern use of working donkeys and cattle in sub-Saharan Africa and elsewhere. The author uses close qualitative analysis of nearly 400 published official and NGO development studies of the complex practicalities of adoption of working animals in developing regions worldwide, in particular of the invisible and under-appreciated donkey. This material, little-used as yet in Ancient Near Eastern archaeology, sheds light on the day-to-day practicalities of working-animal adoption and management – breeding, training, husbandry, hiring and lending. While archaeology will always have need of large-scale anthropological models, the author argues for a parallel bottom-up ethological approach, envisaging the 4th and 3rd millennia BC in Mesopotamia from a viewpoint explicitly acknowledging the major presence of working animals and their daily impact on human activity and the consequent archaeological record.

This innovatory investigation of the role and impact of the donkey in the Ancient Near East and today is an essential handbook for Ancient Near Eastern archaeology and zooarchaeology researchers and students, as well as historians, anthropologists and ethnographers examining the impact of working animals on past and present societies. Wider audiences include the growing sector of human-animal relationship studies, and NGOs concerned with the use of working donkeys worldwide.

Jill Goulder, after an earlier career in industrial marketing research, earned a PhD at the Institute of Archaeology UCL in 2018, employing modern development study material from Africa and elsewhere to provide insights into the early systematic use of working animals in Mesopotamia. See www.jgoulder.com/archaeology.

UCL Institute of Archaeology Publications

Archaeology of African Plant Use
Edited by Chris J. Stevens, Sam Nixon, Mary Anne Murray, and Dorian Q. Fuller

Mediterranean Voyages
Helen Dawson

Wrapping and Unwrapping Material Culture
Edited by Susanna Harris and Laurence Douny

Plain Pottery Traditions of the Eastern Mediterranean and Near East
Edited by Claudia Glatz

Ethnic Ambiguity and the African Past
Edited by Francois G. Richard and Kevin C. MacDonald

Reanimating Industrial Spaces
Edited by Hilary Orange

Living in a Landscape of Scarcity
Laurence Douny

Divine Fertility
The Continuity in Transformation of an Ideology of Sacred Kinship in Northeast Africa
Sada Mire

Working Donkeys in 4th-3rd Millennium BC Mesopotamia
Insights from Modern Development Studies
Jill Goulder

For more information on this series, please visit: www.routledge.com/UCL-Institute-of-Archaeology-Publications/book-series/UCL

Working Donkeys in 4th-3rd Millennium BC Mesopotamia

Insights from Modern Development Studies

Jill Goulder

Routledge
Taylor & Francis Group
LONDON AND NEW YORK

First published 2020
by Routledge
2 Park Square, Milton Park, Abingdon, Oxon OX14 4RN

and by Routledge
52 Vanderbilt Avenue, New York, NY 10017

Routledge is an imprint of the Taylor & Francis Group, an informa business

First issued in paperback 2021

British Library Cataloguing-in-Publication Data
A catalogue record for this book is available from the British Library

Library of Congress Cataloging-in-Publication Data
Names: Goulder, Jill, author.
Title: Working donkeys in 4th–3rd millennium BC Mesopotamia : insights from modern development studies / Jill Goulder.
Description: Milton Park, Abingdon, Oxon ; New York : Routledge, [2020] | Series: UCL institute of archaeology publications | Includes bibliographical references and index.
Identifiers: LCCN 2019040643 (print) | LCCN 2019040644 (ebook) | ISBN 9780367422042 (hardback) | ISBN 9780367822682 (ebook)
Subjects: LCSH: Donkeys—Iraq—History—To 634. | Working animals—Iraq—History—To 634. | Domestication—Iraq—History—To 634. | Human-animal relationships—Iraq—History—To 634.
Classification: LCC SF361.3.I72 G68 2020 (print) | LCC SF361.3.I72 (ebook) | DDC 636.1/820935—dc23
LC record available at https://lccn.loc.gov/2019040643
LC ebook record available at https://lccn.loc.gov/2019040644

ISBN: 978-0-367-42204-2 (hbk)
ISBN: 978-1-03-208381-0 (pbk)
ISBN: 978-0-367-82268-2 (ebk)

Typeset in Sabon
by Apex CoVantage, LLC

This book is dedicated to my father (1918–1981), who hated equids but who would have liked this

Contents

Figures

Tables

Acknowledgements

I should like to thank warmly many people, including the following.

My valued doctoral thesis supervisors at the Institute of Archaeology UCL, Dr. Louise Martin, Professor David Wengrow and Professor Stephen Shennan, and my 'honorary supervisors' Professor William Clarence-Smith of SOAS and Dr. Harriet Crawford of UCL, who helped me out of holes into which I fell. At the Institute of Archaeology, I would particularly like to thank Professors Ruth Whitehouse, Sue Hamilton and Dr. Marion Cutting for their support, as well as the anonymous academic reviewers, and at Routledge Matt Gibbons, Kangan, Louise, Deborah and all the production team. I also fully acknowledge the role of my PhD examiners Professor Paul Halstead of the University of Sheffield and Dr. Mark Altaweel of the Institute of Archaeology.

The late Professor Andrew Sherratt, whom I was privileged to call my friend and upon whose shoulders I hope to stand.

A range of 'donkey-friends' including Professor Paul Starkey, Dr. Peta Jones, Ed Emery of the Donkey Conference, and Professor Haskel and Dr. Tina Greenfield. Also Dr. Carol Palmer at the British Institute in Amman, Dr. Jaafar Jotheri, Dr. Abubakar Suleiman, Nasser Kalawoun and Melissa Upjohn.

For my Burkina Faso research visit in 2013, my friend Bettie Petith of FITIL and in Burkina Faso my friends and colleagues Robert Ilboudo, Bako Bazoin and Alphonse Hien. For my Ethiopia research visit in 2014, Stephen Blakeway, and the Donkey Sanctuary team in Addis Ababa and Bahir Dar, particularly Dr. Bojia Endebu, Dr. Asmamaw Beyene, Anteneh Kassa and Dr. Tewodros Tesfaye Mekonnen.

For both research visits, I would like to thank the UCL Doctoral School and the Institute of Archaeology for their important contributions to the funding of my research visits, and additionally the British Institute in Eastern Africa for their generous grant towards funding my Ethiopian visit.

On illustrations, kind help has been given in particular by Harriet Pottinger and others at the Donkey Sanctuary, The Brooke, Sue Sherratt, Ronika Power, Frank Förster and Dr. Rudolph Kuper, Rick Hauser, Dan Potts, the late Lamia al-Gailani, Ianir Milevski, Ann Searight, Noah Snyder-Mackler, Galen Frysinger, Bruno at http://theonearmedcrab.com and HarperCollins.

Neil Thomson and John Downie gave me invaluable technical help on image resolution, line-drawings and other illustration difficulties.

Some of the photos of present-day working-animal use in this book are of imperfect quality, included nevertheless because they valuably record fast-disappearing lifeways and practices. This applies in particular to my own photos from my fieldwork trips to Burkina Faso and Ethiopia, at a time when I had no idea that a book would result; I apologise for all their failings.

Preface

The arguments and information in this book are strongly based on analysis of published material from more than 1,000 modern development studies and archaeological and other sources. For practicality I have had to place severe constraints on the number actually cited: I ask readers to accept that there are further good references to many of the points that I make.

I provide two resources for further scholarly investigation of the modern development studies in particular: in Appendix B I list examples of key volumes (workshop proceedings, resource books, etc.) of articles and papers on aspects of modern use of working animals as investigated by NGOs and official bodies, and at http://www.jgoulder.com/donkeys. I provide a full compendium of references to all points referred to in this book with, where possible, URLs for the modern development studies.

I use the following broad Mesopotamian chronology:

Table A Mesopotamian chronology

BC (cal.)	*Period*	*Notes*
(4400)4200–3900	Late Ubaid	In S. Levant, Mid/Late Chalcolithic 4200–3200; in Egypt, Naqada I 4000–3500
3900–3600	Early Uruk	In Egypt, Naqada II 3800–3300
3600–3400	Mid-Uruk	In S. Levant, EBIA (3700)3800–3300
3400–3200	Late Uruk	Eanna IVa pictographs; in Egypt, Naqada III 3300–3100
3200 (3000)–3100 (2900)	Jemdet Nasr	Archaic Texts (and late Uruk); in S. Levant, EBI to (3100)2900
3100–2900	Early Dynastic I	In Egypt, Early Dynastic 3100–2650
2900–c.2600	Early Dynastic II	Mari I; Ebla; Ur; in S. Levant, EBII/III
c.2600–2350	Early Dynastic III	Ur burials; Abu Salabikh; Ebla; in S. Levant, EBII/III; in Egypt, Old Kingdom 2675–2130

BC (cal.)	*Period*	*Notes*
2350–2150	Old Akkadian	Sargon; Beydar
2300(2112)–2000	Ur III	Akkad, Umma, Lagash/Girsu, Nippur, Isin, Larsa, Adab, Drehem, Gudea
2000(1900)–1600(1800)	Old Assyrian	Kaneš; Hammurabi
(2000,1900) 1830–1759 (1600)	Old Babylonian	Mari II, Amorites
1700–1500	Early Assyrian	*Farmer's Instructions*; Kassite
1450–1000	Middle Assyrian	

1 Working donkeys then and now

Approaches and contexts

1.1 A new approach to study of working animals in antiquity

From the first systematic domestications of equids for work until very recent times, a key means of land transport has been working animals. A hundred years ago many of us in the West, and our materials and goods, would have travelled by horse. Yet in the developed world the close daily presence of working animals (with their attendant installations and activities) has been rapidly wiped from our consciousness. A different path of consciousness occurs for agricultural use of animals, with their complex and varied uses becoming ossified into an abstract picture of "ox-ploughing" as an economic benefit, in a form of mediaeval Western meme.

My attention was drawn to this in the process of my research into clues to 4th-3rd millennium BC Mesopotamian social and economic daily life at a time of urbanisation and growing social complexity. Stemming from Sherratt's seminal secondary products model (see later), I grew increasingly interested in the presence of working animals in the landscape and in related human-animal relationships, and so in means of investigating these such as examining the use and impacts of working animals worldwide today.

Analogy with modern working-animal use

I have taken a new interdisciplinary approach, addressing the subject of the daily practicalities of working-animal usage through detailed qualitative analysis of the large body of published official and NGO studies of working-animal use in developing areas today. I have focused in particular on sub-Saharan Africa where, in many regions, working cattle and donkeys have only been adopted in recent decades. My aim has not been comprehensiveness, but a well-based large, representative sample of key accounts of modern use of working animals, particularly those involving societies previously unfamiliar with any form of power except that of humans for agriculture and transport. As part of my research I also undertook brief visits to rural areas of Burkina Faso and Ethiopia, in 2013 and 2014, to gain appreciation of the circumstances and cultures in which such studies are set; I refer to my personal observations from these visits as 'JG Burkina Faso 2013' and 'JG Ethiopia 2014'.

There has been only very limited archaeological use of modern working-animal study sources in the large body of Ancient Near Eastern studies to date. Ethnoarchaeology relating to animals has to date focused largely on the impact on human society of hunted and herded animals, with very little attention to the phenomenon of working animals. Renger, in 1990, consulted official handbooks on draught animal use in sub-Saharan Africa for clarification of the training and husbandry of working animals in later 3rd- and early 2nd-millennium BC Mesopotamian texts. Bogucki (1993), a key post-Sherratt source for insight on the implications of the adoption of working animals in early Europe, cites the African work of Barrett *et al.* (1982; a large survey in 27 villages in NE Burkina Faso), Singh (1988; a large qualitative survey among farming families in seven villages in the very different environment of central Burkina Faso), and McCann (1984) and Peters (1986; overviews by experts in Ethiopia and Botswana), to support arguments on the economics of working-animal use in relation to labour. Brodie (2008), focusing on the Aegean, cites four similar African sources, and Goody (1976) refers to several of the same genre as examples of his wider findings; Halstead (e.g. 2014) makes use of many years of informal ethnographic observation among traditional farmers in present-day Greece.

Such modern studies offer a largely untapped resource for assessment (with caveats) of the likely on-the-ground role and impact of working animals in antiquity – in particular on the unexamined role of donkeys. They investigate at farm and household level the practical social and economic ramifications of their adoption and use, and they demonstrate vividly the complexities of working-animal operational systems, often in situations where the facilities today include little that was not available in the 4th millennium BC.

Addressing Mesopotamia and working animals

Archaeological thinking about 4th-3rd millennium BC Mesopotamian economy and society has been dominated by large-scale themes since Gordon Childe's defining in the 1930s of the Urban Revolution. From the 1980s, discussions of the 'Uruk expansion' by Guillermo Algaze and others have taken this region and period as a platform for the introduction of world systems theory into the archaeology of the Middle East. The emphasis has remained on macro-scale patterns with a particular focus on long-distance trade patterns and on the advent of (ox-)ploughing and the implications for urbanisation. In this book I am instead approaching changes in human-animal relations at the micro scale, from the bottom up, including both basic logistical matters of breeding, supply, training, grazing and foddering, the impact of working animals on local economies through processes such as labour adjustments, hiring and lending, and the central contribution of short-distance transportation work. A key objective has been to challenge existing archaeological assumptions – with an only fragmentary evidential base – about the role and on-the-ground impact of these animals in early complex societies, offering foundations for corresponding implications for higher-level social and economic theory. The complex minutiae of daily life with working animals – seeking ways of using a ploughing ox year-round, keeping a multi-purpose working cow, developing new income sources and household labour systems from donkey ownership – build up to a series of model adaptations hardly proposed yet.

While I am using the example of late 4th- and 3rd-millennium BC Mesopotamia in order to draw upon my specialist knowledge, there is wider applicability of my approach to other Ancient Near Eastern and European cultures featuring early systematic working-animal use. I argue the widespread need for a new working-animal mindedness among fieldworkers and commentators in the Ancient Near East and beyond, based on practical understanding of the likely traces of their presence and impact.

Invisible donkeys

Aside from what I argue as general 'donkey-blindness' in high-level models, zooarchaeological detection of working donkey use can be very problematic. Working donkeys in many cultures, from antiquity to today, have surprisingly rarely been eaten, with the carcasses left in the desert or dragged outside settlements, so they are very under-represented in food-middens. In earlier 20th-century AD Mesopotamian archaeology, too, investigation was largely settlement-oriented; so again, any working donkeys in the small farms and villages were not detected. To add to their invisibility, *Equus asinus* remains are notoriously hard to tell from those of the onagers (*Equus hemionus*) native to Mesopotamia; these were commonly hunted for meat and hides, but never domesticated for work, though there was use of them in the 3rd millennium BC for hybridising with donkeys. These often-unconsidered aspects are of central importance to my case for the under-estimation of the role of donkeys in Mesopotamia, and I have included some notes on this aspect in Chapter 2.3.

Where it all began: Sherratt's secondary-products model

Following long-running scholarly debate on the origins and dating of early ploughing and wheel use, Sherratt (e.g. 1981, 1997, 2003, 2006) led working animals into the centre of the prehistoric archaeological arena with his far-reaching theory concerning the advent of secondary products – traction, milk and wool – in the 4th millennium BC; Figure 1.1 shows his model of working-animal usage. He originally argued that these products were evolved or adopted as a package, including oxen (castrated male cattle – see the Glossary in Appendix A) for ploughing and carts and with a distinct but

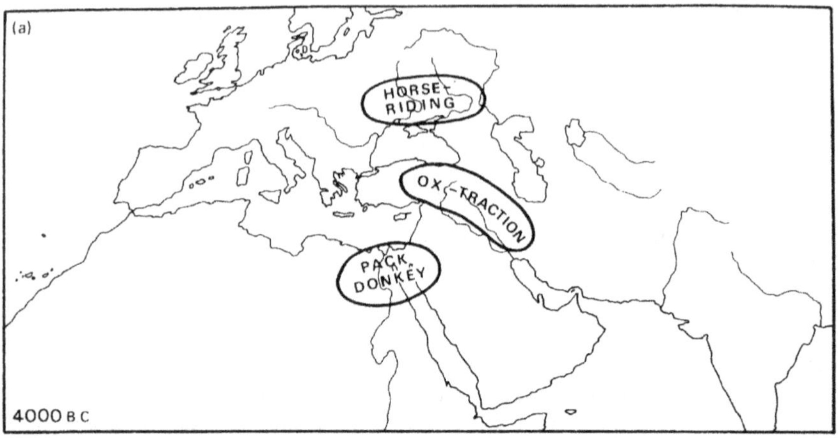

Figure 1.1 Sherratt's (1981:288) ox-centric view of early working-animal use in Mesopotamia

Source: By kind permission of Sue Sherratt

roughly contemporary enlistment of equids for long-distance pack transport (Sherratt 1981:263/295). Sherratt (1981:262/299) also drew social evolutionary elements from Goody's (1976) analysis of Murdock's (1967) *Ethnographic Atlas*, linking cattle-herding and ox-ploughing with new land-based inheritance structures and with the exclusion of women from economic and social power.

Sherratt's model focused on the transmission of specific hardware concepts – the plough and the cart – between the Near East, Europe and the Central Asian steppe in approximately the 4th millennium BC. He took the view that although localised adoption of ploughing with cattle took place in Mesopotamia from perhaps the 6th millennium BC, with some possible earlier use for threshing (Sherratt 2006:342–3), the systematic use of yoked oxen for ploughing emerged rapidly in the 4th millennium BC, becoming closely associated with the later Uruk as an enabler of centrally controlled production of an irrigation-based agricultural surplus upon which the new social complexity depended for its survival (e.g. Sherratt 1981:290–8, 2006:343).

Archaeological thought now – while acknowledging the highly significant impact on human society of the use of working animals – generally takes the view that there was not a moment of innovation or a package involving the close interrelation of several secondary products and traction modes diffusing as a unit. Vigne and Helmer (2007:34–5) consider that Sherratt initially undervalued the technical ingenuity of earlier cultures, suggesting that a more intricate evolution of human-animal relationships took place in which cattle were managed early on for milk and even work, with hunting still the primary source of meat.

Sherratt's traction complex focused on cattle, linking them firmly with the plough and the cart, with donkeys initially mentioned only as working on the southern Levant/Egypt pack routes (though with later acknowledgement of their spread to Mesopotamia (e.g. Sherratt 2003:238)). Sherratt (2006:351–2) does later comment on the different evolution of the plough itself in the damp, heavy soils of northern Europe compared with those of Mesopotamia and Egypt, but not on corresponding differences in animal use apart from a brief reference (2006:336) to the unsuitability of (female) cows for traction in 4th-millennium BC Europe.

The sign for a plough is seen in the earliest texts in the late 4th millennium BC. The initial use of oxen is, though, more assumed than indicated: Englund (1995:33) points out that in the earliest texts

> [o]nly several uncertain accounts register together the existence of both the plow represented by the sign APIN and oxen represented by the sign GU_4. Whether oxen played a large role in field work in the Late Uruk period is thus a matter of conjecture.

While later official Mesopotamian texts and depictions, recording state-controlled agricultural activities, do commonly refer to oxen, donkeys are nevertheless mentioned as regularly used for ploughing from the 3rd millennium BC, and commentators are now highlighting the evidence of use of cows for ploughing, as in Egypt where cows have been common ploughing animals from antiquity to the present day. In the Aegean, cattle skeleton pathologies at Knossos indicate that cows were used for ploughing there from at least the 6th millennium BC (e.g. Isaakidou 2011:101–3).

Sherratt initially included the horse in his model (1981:272–4), suggesting a steppe 'package' of horse and oxcart (the former for herding, the latter for transport of households to follow the herds) and this may well have influenced his focus on oxcarts rather than pack donkeys for transport in Mesopotamia. (He later (1997:31) withdrew horses from his scenario.) The lack of 'donkey-mindedness' in discussion of this period may have stemmed from historical species availability in the home regions of archaeological commentators, with donkeys not a feature of northern Europe and a prevailing view of the virtues of deep (ox-)ploughing: this has led to 'donkey-blind' models of early working-animal systems. For example, Russell (2012:208), in her otherwise useful text-book on humans and animals in prehistory, inexplicably omits the donkey from her table of 'major domestic animals', and (p. 228) lists only cattle, horses, camels and even possibly dogs as likely early pack animals. In Chapter 6.3 I describe earlier dismissal by commentators of the importance to Mesopotamia of donkey caravans, though now increasingly replaced by recognition of their key role alongside boat transport. These cumulative gaps and biases have resulted in a profile in the Ancient Near East at odds with donkey use in modern, developing regions. Bogucki (1993:492) comments about archaeological approaches to working animals that

> there is a longstanding bias towards viewing livestock as subsistence resources, with far less consideration given to their economic roles while living. This bias is understandable, for faunal remains are archaeological data and prehistoric living animals are abstract concepts. While the concept of 'secondary products' has addressed this issue to some degree, there is still a tendency to think in terms of 'products' rather than viewing animals as 'assets' in the Neolithic economy.

Finally, wheeled vehicles, as epitomised by the famous Uruk-period pictographs (many references, e.g. Littauer and Crouwel 1979:13), have also been in the past an emblematic feature of the Uruk phenomenon; local transportation, if mentioned at all in archaeological commentary, is also still too often casually associated with the wheel. Emberling and Minc (2016:821) comment that

> The first representations of wheeled vehicles also appeared during this time, but as networks of roads would only be fully developed later (as the "hollow ways" of third millennium BCE Upper Mesopotamia suggest – Wilkinson, 1993; Ur, 2003), their significance for Uruk period transportation was likely limited.

Sherratt himself (2003:243/247) subsequently concludes that wheeled vehicles at this early period are principally for ritual or prestige use, with pack or dragging being a more cost-effective and practical means of transport in daily life.

Structure of my narrative

My emphasis in this book is unashamedly on the working donkey, as the often-ignored animal. As I underline throughout, discussion of Mesopotamian working-animal use has, to date, focused very largely on working (male) cattle, and my aim is to redress this culpable imbalance, though I also necessarily examine the case of cattle as important complementary and competing working animals. There is regrettably little data available on working female cows from either modern or ancient sources – another 'invisible' resource which might otherwise have taken a larger role in this book in relation to the economic and practical implications of their use.

My approach is driven throughout by the modern data. My analysis was conducted with a very open mind, respecting the novel nature of my dataset and the need throughout for modern-ancient linkages, so my narrative structure is purposely *post hoc*. Where I have found no modern material in the development studies for analogy with antiquity (e.g. patterns of movement of working animals, diachronic studies), or insufficient clues from antiquity to relate to modern findings (e.g. the use of working animals by women), I do not pursue the subject. Based on the findings from my large data-set on modern use of working animals, I have searched among key commentators on Mesopotamia and sometimes elsewhere in the Near East for illustrative examples of relevant textual, representational, zooarchaeological or other archaeological parallels, my argument being that material from these latter is at best fragmentary and will benefit from the addition of modern material to complement the ancient evidence and shed light on the realities of working-animal adoption in this period.

Rather than aiming to develop a formal new model through such a heuristic approach, my primary goal is to provide 'a guide to the questions we should be asking about the past' (Halstead 1987:77) in the field of working-animal use for transport and traction, through providing groundwork for further study and demonstrating to commentators on 4th- and 3rd-millennium BC Mesopotamia and more widely the value of a new resource which populates the Mesopotamian landscape with working animals and with the practical day-to-day activities arising from their adoption.

The present chapter provides the backdrop for my main arguments and my interdisciplinary approach to the little-addressed issue of the day-to-day practicalities of working-animal use in antiquity. This is followed by a description in Chapter 1.2 of the modern studies that I have used and their origin, region and import, with a note on the history of working animals in sub-Saharan Africa and a brief discussion of the validity of ethnographic studies for archaeological interpretation.

In Chapter 2.1 I give brief notes on the larger-scale political background of 4th-3rd millennium BC Mesopotamia, notably on recent reassessments of established dichotomous thinking about sedentary cultures and the powerful mobile pastoralist groups now thought to be significantly instrumental in urbanisation processes of the period. A related subject discussed here is the growing recognition of rural Mesopotamian working-animal use outside central control and barely recorded in official texts and representations. In Chapter 2.2 I diverge briefly into observations on the likely early domestication process of donkeys – markedly different from that of cattle – and on their curious simultaneous positioning since their earliest use as both invaluable and despised. Chapter 2.3 addresses zooarchaeological issues central to the invisibility of donkeys in the early periods of their systematic use: the misleading scarcity of donkey remains in urban food-middens, and the difficulties of distinguishing between *Equus asinus* remains and those of the Mesopotamian onager (*E. hemionus*) hunted for meat and hides but not domesticated.

In Chapter 3.1 I explain the important behavioural and physiological characteristics of donkeys and cattle which drive the very different social and agricultural trajectories of their adoption and systematic use, as well as noting the role of cultural and environmental obstacles. I also discuss, in Chapter 3.2, the limited material on the insufficiently mentioned wide usage today of (female) cows for ploughing and the implications for antiquity.

Chapters 4–7 form the main body of my findings from the modern development studies. In Chapters 4 and 5 I give an account of the often-unmentioned practical implications of working-animal adoption, which can be major causes of use of labour and land: breeding, hiring and lending, husbandry and training (underlining the key role in working-animal choice of person-to-person dissemination), and grazing and supply of fodder. I include an outline of donkey 'products' – hides, milk, dung, and the important subject of the comparative rarity throughout history of eating of donkey meat, leading to the archaeological scarcity of their remains. In Chapter 6.1 I discuss in detail the unappreciated but crucial role of the donkey in antiquity for short-distance transport, with a short commentary in Chapter 6.2 on other uses of working animals, and a note on selected other traction and transport species. In Chapter 6.3 I expound the striking scarcity of material on long-distance pack-donkey caravans.

In Chapter 7 I focus on the important but often misinterpreted history and practicalities of ploughing, from 4th-millennium BC Mesopotamia to the present day. In Chapter 7.3 in particular I comment from a practical basis on Ester Boserup's anthropological model of plough adoption and the consequences for yield and labour, and I underline key modern findings such as the importance of year-round utilisation. Chapter 7.4 briefly addresses Engels', Goody's and other models of the impact of plough adoption on social inequality, notably the issue of working animals and the role of women.

In Chapter 8 I highlight findings from my detailed analyses that illustrate how insights from modern development studies can provide significant new depth to archaeological investigation, of 4th-3rd millennium BC Mesopotamia and more broadly. They demonstrate for Ancient Near Eastern archaeologists the overwhelmingly likely presence of a range of day-to-day activities and social levelling adjustments closely related to use of donkeys and other working animals, which are barely considered in social and economic models of working-animal adoption.

The unrecognised role of the donkey in particular, as I underline throughout this book, is brought to the foreground, breaking out from the abstraction of 'oxen' and the need for deep ploughing, and highlighting the central role of short-distance (donkey) transport and its social benefits for disenfranchised groups.

Subjects not covered

My focus is on subjects with a reasonable basis for broad comparison between my data-set of modern study analogical material and the possible scenario in 4th- and 3rd-millennium BC Mesopotamia. For this reason – notably relating to modern use of metal – I am excluding all but a brief overview of wheeled vehicles, sledges and travois, stationary power, harness, and competing or complementary transport systems such as boats.

A particular disappointment has been the largely insuperable difficulty of the scarcity of material on women in ancient Mesopotamia for comparison with the rich data on the transformative role of donkey transportation use for women in developing countries today. I discuss briefly in Chapter 7.4 the issue of cultural limitations for women on ox-use for ploughing in modern developing regions and, in Chapter 6.1, in more detail the new freedoms conferred upon them by adoption of donkeys for local transportation.

Even more scarce, both in antiquity and today, is material relating to the detailed operation of long-distance donkey caravans. Long-distance caravans are often deliberately invisible, from prehistory to the present day, the caravaneers preferring to stay under the radar of authorities and attackers, and nowadays also of the attentions of animal-welfare NGOs. On-the-road evidence in 4th-3rd millennium BC Mesopotamia lies outside archaeological fieldwork reach to date, despite advances in satellite imagery. My chapter on this subject relies notably on anecdotal accounts of travellers in recent history encountering such caravans.

Onager-donkey hybrids were employed for work in the 3rd millennium BC (see my Glossary on onagers in Appendix A), but I do not discuss them in any detail as they have no truly comparable modern equivalents. Horses and camels were effectively absent from Mesopotamia in much of the period under discussion, and their use in much of sub-Saharan Africa is not widespread; for related reasons riding is excluded.

1.2 Modern studies of working animals

A new body of studies: donkeys and other working animals in developing regions

Working-animal use worldwide

The number of working donkeys worldwide today is hard to assess but is estimated at 46 million (Donkey Sanctuary/FAO statistics), the largest population being in Ethiopia. The worldwide working-animal population more generally is equally hard to calculate, but in recent years may be 200–250 million (Starkey 2011:11–12 – a valuable source of information on incidence by region); animal power is estimated to form the main source of energy for agriculture and transport for half the world's population. While much agricultural and transportation power in developing regions is still provided by humans, and plenty by mechanised equipment, working cattle and donkeys are widely used in Africa and in much of Asia and Latin America, and on a much smaller scale in parts of Europe, where in eastern regions, female cows and equids still have a significant working role. The donkey population in some sub-Saharan African regions is increasing, though use is declining overall in most other regions worldwide, with competition from tractors, pick-up trucks, bicycles and light motorcycles.

In the modern Near East direct regional analogy with use of working animals in antiquity would be largely inappropriate. There is major mechanisation, with low usage now of donkeys in Turkey, Jordan and Lebanon, though they are still a transportation staple in regions of Iraq and Yemen. In Egypt donkeys have been common pack animals since antiquity but have seldom been used for ploughing despite suitable soils, and in Greece and Turkey the use of donkeys for ploughing is patchy though known from Classical times; donkeys are used in small numbers for ploughing in Syria, Jordan and Iran today (e.g. Figure 1.2). Modern NGO and official studies on working animals in Mesopotamian regions are scarce – particularly outside Arabic-language publications (which I am not yet equipped to read) – for a range of reasons related to cultural attitudes to animals and inevitably to the history of damaging conflicts in this region; a welcome exception is Tabbaa's detailed (2003) study of the use and management of donkeys and mules in history and in present-day Syria.

Working-animal development studies

In recent decades as enthusiasm for fossil-fuel technology has waned in developing regions, a new focus by agencies and NGOs has emerged on the value of promoting use of working animals. Animal-based farming and transport

Figure 1.2 Ploughing with a donkey and mule in Jordan

Source: © Carol Palmer 2014, by permission

work takes place with very little realistic statistical recording – a gap which has been filled by numerous small-scale surveys and studies. I have employed a rigorous qualitative analysis approach to the modern studies available, drawing on a previous career in industrial market research consultancy. Figure 1.3 summarises the publication dates of the 389 modern studies used from all regions worldwide, and Table 1.1 their distribution by region (note that some studies cover more than one region and are included against all relevant headings). Note that the incidence of studies by region reflects various factors such as official/NGO focus, stability of government, presence of working animals, and also (for my using) language of publication (mainly English, French and Spanish).

Many of my sources were published in the last 40 years as part of an increasing drive to encourage and monitor the use of working donkeys and cattle in sub-Saharan Africa in particular as a more sustainable alternative to mechanisation. Through a combination of factors there has been no widespread history in this region of use of donkeys and cattle for work; the European colonial history of this region, and its strong candidacy for developed-country aid, has led to the publication of a unique body of official and NGO studies on responses and adaptations to the new animal facility. A fruitful source of studies has been the web-based Animal Traction Network for Eastern and Southern Africa (ATNESA, www.atnesa.org). Workshop papers and studies are published by major working-animal NGOs such as The Brooke and the Donkey Sanctuary and by international official organisations (FAO,

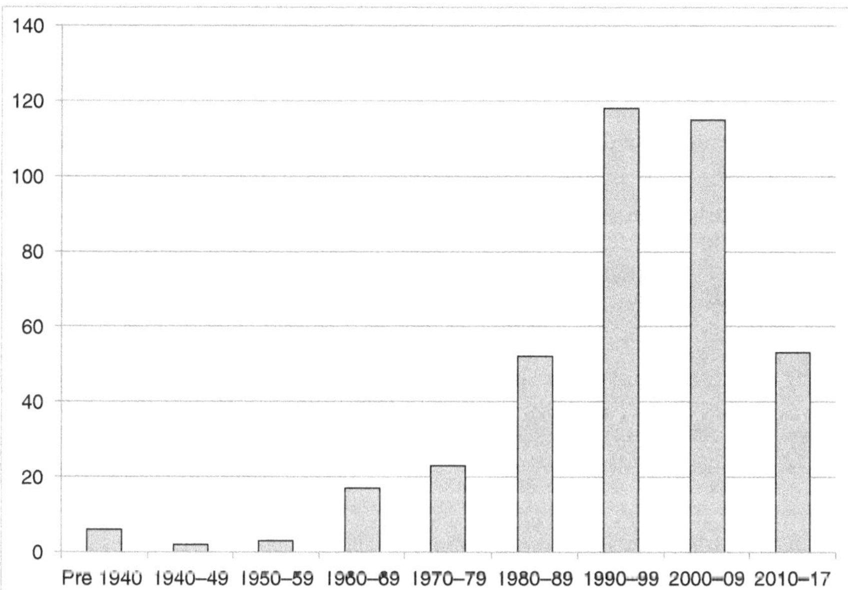

Figure 1.3 Publication dates of the modern/ethnographic studies used (total 389)

Source: © Jill Goulder 2019

Table 1.1 Modern studies used by region worldwide

Region	*No.*
TOTAL	**389**
General/global	14
Sub-Saharan Africa	269
Asia	50
Near East/North Africa	33
Europe/Eurasia	32
North America	10
Latin America	16
Australia	3

ODI, World Bank) and academic institutions. I list some examples of collections of articles and papers (among many more) in Appendix B.

I have retrieved some useful NGO and other studies in East Asia on the employment of working animals, which give important insights into utilisation and husbandry of donkeys and female cows in particular; but unlike much of sub-Saharan Africa, working-animal technology is long-established in the region, so providing fewer potential parallels with the initial periods of systematic use of working donkeys and cattle. In modern Latin and Central America the aid structure is very different; there is only limited published material on the role and activities of working animals, and donkeys are often excluded from working-animal surveys and veterinary college curricula. More generally there are examples worldwide of lack of attention to the donkey by some (urban) official bodies, who have often regarded donkeys as old-fashioned technology not in keeping with their modernising approach.

I have also made use of a large body of published practical experimentation and specialist commentary on the biological capabilities and constraints of donkeys in particular in relation to pack and traction, including work rates and capabilities, and food/water needs and preferences. Also of value are studies of the social behaviour of donkeys in the wild, relating to their propensity for taming and handling, pack-carrying and traction activity, and breeding in captivity.

Finally, looking into history, there are 19th- and 20th-century AD works by explorers and anthropologists which include on-the-ground descriptions of working-animal use by traditional farmers, nomadic groups and pack-caravaneers.

The development studies from sub-Saharan Africa

I concentrate in particular on the material from sub-Saharan Africa (Figure 1.4 shows the geographical distribution of the studies used), which has

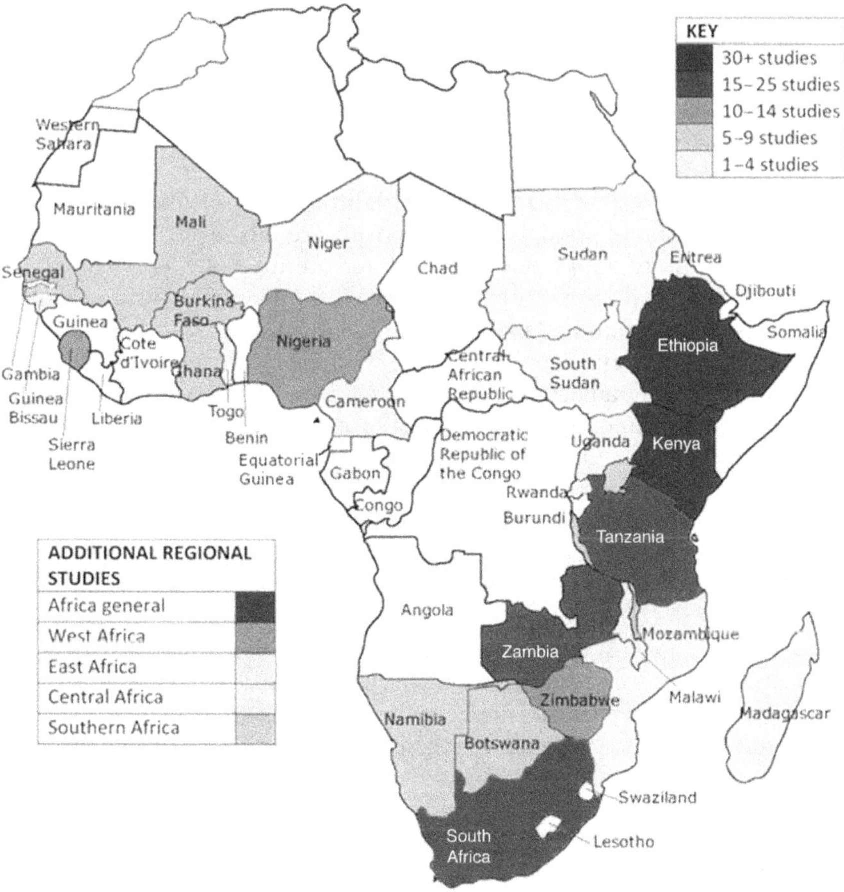

Figure 1.4 Distribution of modern studies used in sub-Saharan Africa (total 269)
Source: © Jill Goulder 2019

the twin advantages of rich new study data and an often relatively pristine environment in which to study the first transitions from hoe agriculture to animal traction and from human porterage to pack animal with no intermediate use of mechanical means.

For reasons of space in this book, I am constrained from referencing a good proportion of the 389 modern development studies which I have consulted, though I have listed and described them all for further consultation at http://www.jgoulder.com/donkeys and given example resources in Appendix B.

Through very detailed reading of these studies, I have extracted a large harvest of themes and categories of findings about the daily actualities of using working animals, and on the impacts at farm-, household- and village-level of their adoption.

I discuss in what follows the perils of ethnographic analogy: the findings of modern studies should of course not be accepted unconditionally. African animal traction studies may range from anthropological modelling to accounts by agricultural engineers and agronomists, the latter at times based on trials at research stations rather than in the field; their agenda, sometimes devised by urban-based official agents with little on-the-ground knowledge, is the promulgation of possibly isolated positive results in a drive to encourage local farmers to adopt new 'modern' agricultural systems. Meanwhile, NGO-commissioned studies of the use of working animals in developing regions specifically address social and economic aspects, but are likely to pitch their sampling and reporting in line with their worthy aim of improving conditions for animals and encouraging donations, with the potential result of polarising findings into pre-intervention (bad) and post-intervention (good). With suitable filters, though, I argue that both categories of study can provide valuable material for the examination of situations in antiquity. My intention has been to focus ethologically (see what follows) on irreducibles relating to the actuality of the animals and their daily management – something largely missing in model-driven commentary – and on arguing for recognition of the often complex nature of the decision paths that individuals and cultural groups take in relation to working-animal usage.

Soil and climate are of course among the many factors affecting choice and usage of working animals, and I refer to these where relevant as an analogical constraint. My data-set of modern studies would certainly not be appropriate for systematic critical evaluation of these particular factors, and I am not attempting this: most of the studies are very small-scale, pinpricks in the often micro-varied ecological landscape, and in the cases of sub-Saharan Africa, the animals discussed have often initially been introduced to the area artificially. There is wide evidence that both cattle and donkeys can, in practice, be used in most environments, preference being guided by other factors; only in extreme ecological zones (severe drought or very heavy damp soil) does environment override other factors.

Recent history of working animals in sub-Saharan Africa

In many regions of sub-Saharan Africa the use of working cattle and donkeys has, remarkably, only been introduced within the last 100 or even 50 years. Cattle (initially taurine) were domesticated for meat, milk and prestige by pastoralists in antiquity; there are some accounts of early modern use of oxen for pack and for lifting water. From the 1st millennium BC, donkeys were brought down into East Africa from the north, attested in small numbers in Tanzania and in Kenyan pastoral herding contexts. Donkeys continued to filter south into sub-Saharan Africa in pack caravans, and by the

19th century AD donkeys were also being brought in by boat from northeast Africa, the Gulf and India (Clarence-Smith 2016:112).

Joubert (1995:126/135) records the introduction of cattle and donkey traction and donkey pack into South Africa by European settlers in the 17th century AD, with new waves of introduction to some areas from the 19th century AD, but in most sub-Saharan regions, cattle and donkey use for work remained rare. Wheeled transportation was scarce in pre-colonial Africa, and horses were largely used for riding by elites.

Factors in non-adoption

During the 20th century AD awareness among farmers of animal-traction possibilities must inevitably have increased, but adoption largely waited for specific colonial or other external initiatives. A range of commentators (notably Clutton-Brock 2012:109–16; Goody 1976:25–33; Kjaerby 1983; Münzinger 1982; Starkey 1994a:71) suggest that factors in the widespread non-adoption of animals for traction may have included the lack of suitable animals, endemic animal disease in some regions, cultural prejudices, absence of profit motive, wide separation of arable farming from nomadic livestock-keeping (see Chapter 2.2 on this relationship), shortage of land/abundance of land, shortage of water/grazing, unsuitable soils/terrain, agriculture based on shifting cultivation (Figure 1.5), and existing labour systems (including slavery, up to the 20th century AD).

Figure 1.5 Woman (with baby) hand-cultivating near Ouagadougou, Burkina Faso. Note the donkey under the tree

Source: © JG Burkina Faso 2013

Colonial and post-colonial working-animal initiatives

In the 20th century AD, under various colonial regimes, draught cattle were introduced into a number of regions, driven by a new emphasis on cash crops such as cotton and peanut. Both early and later initiatives often faltered or failed as a result of poor research, planning and cultural understanding: equipment was heavy and inappropriate, cattle died from trypanosomiasis (see the Glossary in Appendix A), additional factors such as labour for other activities were not taken into account, too many changes were introduced at once, and adoption was hampered in some regions by heavy-handed regulation on local groups to protect the income of European settlers.

In some regions these productivity-led initiatives then developed post-war into an emphasis on mechanisation, and this intensified in the post-colonial era as new governments worked to establish their modernising credentials (see, among many others, Havard *et al.* 2007:28; Panin and Ellis-Jones 1994:94; Pingali *et al.* 1987); it took the 1973 oil crisis to bring home the inflexibility of developed-world, combustion-engine-driven solutions to African agricultural development. Cattle traction initiatives then started or revived in the 1970s and 1980s, both in regions where tractors and/or animal traction had been tried and in areas without experience beyond hoe agriculture. By the start of the 21st century AD hoe agriculture still predominated in sub-Saharan Africa but – as a result of the new mindsets – animal traction had become significantly more widely used than mechanical means. Economically-challenged government authorities in various regions have now been reducing economic support for animal traction, with challenges being addressed instead by NGOs and the private sector. Tibbs (1989:3) alerts us to the attitude of organisations such as the World Bank, who in 1987 explicitly withheld research funding '"[b]ecause of the simplicity of animal draft technology"' which, therefore, was not perceived as requiring it.

Official working-animal drives and subsequent local adaptations

Meanwhile, initiatives by authorities still tended to distort natural adoption and expansion of the most locally suitable agricultural and rural transport systems and to ignore established local practices such as use of donkeys. Adopters of working animals for ploughing through official/NGO initiatives may, in practice, take more interest in their use for farm transport or for income-earning transportation for others, particularly in the case of donkeys. Transport activity generally heightened with the advent of ploughing animals; Starkey (1994a:75) reports how some African farmers, having been encouraged to invest in animals for ploughing, return to manual cultivation and use their animals instead for profitable work transporting for others. Authorities focusing on cash crops and a specific agricultural model made vain attempts to stem this shift: in Ivory Coast at one point, oxcarts were even not issued to farmers because their use would divert cattle from use in cotton cultivation (Landais and Lhoste 1990:222).

In modern societies, too, donkeys are often insufficiently acknowledged as valuable work resources. There are striking African examples of suspicion among government authorities of donkeys, expressed in misleading and denigratory information disseminated by urban officials (Pearson *et al.* 2001:81 and others) and in ill-conceived culls, as vividly described in Jacobs' 2001 article *The Great Bophuthatswana Donkey Massacre*. Donkeys are often wrongly seen by external bodies as a poverty indicator (Cochin 1995:55) rather than an engine for change. Another element is the social assignment by men – in many cultures the only users of working cattle – of donkeys to working use by women, who are often debarred for apparently cultural (but possibly acquisitive or power-oriented) reasons from using working oxen, but this very often applies less to donkeys.

There has been a good deal of development literature in recent years about the complex and unexpected paths that working-animal use is taking now that the focus on imposition of northern European models – with emphasis on investment in a pair of oxen and heavy-duty equipment – has lessened. Although in some areas a drive to modern mechanical options persists, small-scale farmers, increasingly on their own initiative and through word of mouth, have switched to single-animal donkey or cow draught and lower-power implements (e.g. Vall *et al.* 2002:120 on Cameroon). Donkeys are often more suitable than cattle for ploughing in areas with light, sandy soil, aridity and heat, disease and parasites, and with scarcity of grazing, fodder or water (see earlier). They constitute a valuable low-cost option for secondary tillage and light agricultural operations (Table 3.1) – a feature perhaps apparent in 3rd-millennium BC Mesopotamian records indicating cattle and donkeys ploughing but with the donkeys additionally working on seeding operations (Zarins 2014:190). Donkeys in modern usage also 'relieve rather than exacerbate the harvest bottleneck' (Brodie 2008:303) through their transport work, and as I underline throughout, are a highly flexible resource for general year-round use including income-earning work.

The case of Ethiopia

A notable exception to the scarcity of working animals in sub-Saharan Africa until recent times is Ethiopia. This country has a human population of 80–100 million, 7–8 million oxen (Starkey 2011:12/33) and a growing donkey population of nearly 9 million (Donkey Sanctuary pers. comm. 2019). Mechanisation of ploughing and transport activities is very low, and there is considerable NGO focus on working animals. The 'maresha' wooden ard plough is thought to have been used here for at least 3,000 or even 5,000 years, possibly brought in from Egypt/Arabia or developed independently. Goody (1976:109) suggests that the very early adoption of ploughing reflects the country's character as 'African in a geographical but not in a social sense', with long-established links with North Africa and Western Asia. Meanwhile, although the natural distribution of wild donkeys crosses the northern Ethiopia border, it is considered that donkeys were not initially domesticated here. Egyptian texts from c.2,270 BC report pack-donkey caravans between Egypt and Punt (Ethiopia), but significant use within Ethiopia is not detected before the end of the 1st millennium BC.

Use of ethnography for study of working-animal use in prehistoric cultures

Few of my sources are strictly ethnographic, but rather modern studies unrelated in themselves to analogy with the past; their use comes under a wider umbrella of modern analogy. The use of ethnographies for study of ancient societies is an established approach, valuable in addressing potential biases and gaps by providing dimensions beyond the individual present-day experience that may subconsciously inform our deliberations (Goulder 2016:67–8). Sensitively used, they can explore the range of responses to situations, possible behaviours and interpretations, identifying key variables and assisting in 'eliminating error and assessing likelihood, improving credibility and delimiting uncertainty' (Wylie 2002:145). Shennan (2004) argues that in order to leap beyond mere 'plausibility' (p. 4) in interpreting the actions relating to archaeological assemblages, there is significant value in hearing from 'someone who knows how to do it' (p. 17).

In many of the recent working-animal studies by developed-world and locally based agencies in Africa and elsewhere there has been an explicit intention to move away from northern European priorities and to establish local needs and practices with close attention to suitability of solutions to the local environment and farming traditions. This offers archaeologists the opportunity to reassess the often Western-centric epistemology of early working-animal use in the Near East, where there has been little challenge to assumptions such as that decisions, by animal-users and others, were always directed towards greater utility, productivity and profitability in developed-world terms (Hodder 1982:211; Wylie 2002:145). Kristin Armstrong Oma (2007:1) suggests that study of animals in the human sphere has been caught between 'the cost-benefit thinking of modern industrial farming' and 'symbolically oriented archaeology motivated by anthropology'.

Modern human study subjects are inevitably rarely free from some form of contamination, by Government subsidies, new employment patterns, and interaction with other societies. Ethology – the study of innate animal behaviour and psychological responses to situations – can fulfil a valuable role in rectifying some of the unavoidable shortcomings of ethnographic field interviewing, through observation studies of the operation of animal-based traction and transport.

2 Working animals in antiquity

2.1 The adoption of working animals in the Ancient Near East

Working animals in 4th- and 3rd-millennium BC Mesopotamia

The systematic adoption of animals for work in the 4th and 3rd millennia BC developed in specific social and economic environments including Mesopotamia, the southern Levant, Egypt and Europe. Of these, the former three regions had early access to the donkey – a key element of the 'stable' of working animals that persists to this day. In a full analysis of the advent of working cattle and donkeys all three Near Eastern regions should be represented, but for compactness of argument – and to reflect my own specialist knowledge – I focus chiefly in this book on examination of their impact in one example region, 4th- and 3rd-millennium BC Mesopotamia (Figure 2.1). I give a brief background outline of this period in what follows.

Precursors

Prior to the 4th millennium BC, the limited archaeological evidence available indicates that the Ubaid settlements of southern Mesopotamia, and in contemporary but perhaps separate manifestations in the north, contained the seeds of social stratification and urbanisation but had not coalesced into a state entity. While earlier arguments posited the Ubaid as an immediate precursor to the state development of the 4th-millennium BC Uruk, newer evidence indicates a less direct path (Carter and Philip 2010:8–10). Pournelle (2003:83) revises established thinking on the landscape and agriculture of southern Mesopotamia at this time – typically modelled as irrigation-fed agriculture employing oxen for traction – with her satellite-imagery-based model of an intricate wetland hinterland from which the emerging cities derived a significant part of their resources.

The Uruk phenomenon

The urban development of the Uruk period (c.3900–3200 BC) rested on organisational systems controlling the flow of goods and information, culminating

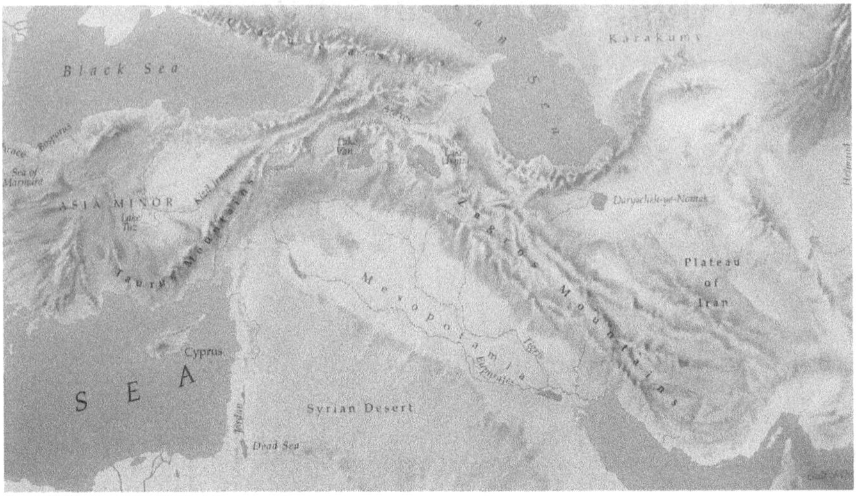

Figure 2.1 Topographical view of the Mesopotamian region

Source: Extract from Times Atlas of the World 1995:43. © Collins Bartholomew 1995, by permission

by the late Uruk in the emergence of cuneiform writing for bureaucratic administration. Control of the rapidly growing urban population was conducted through 'new forms of organizing labor that delivered economies of scale in the production of subsistence and industrial commodities to southern societies' (Algaze 2008:9), including for the management and use of working animals for ploughing and transport. The complexity involved in this new social process generated an elite responsible for the acceleration of demand for monumental architecture and luxury goods, and a lower class of encumbered workers and captives – the latter's social position demonstrated by the similar terminology in the late 4th-millennium BC Uruk Archaic Texts for these and domestic animals (e.g. Tani 1996:403–5).

In the mid-Uruk, a range of outposts and transshipment centres relating to the southern core then developed in north and north-eastern Mesopotamia, often concentrated at east-west transport route intersection points with major rivers and at the base of mountain routes (e.g. Matthews and Fazeli 2004:70) with smaller waystations between.

The northern powerhouse

The marked refocusing of archaeological fieldwork onto northern Mesopotamia in more recent times has led to a reassessment of the development of urbanisation and complexity here, formerly considered to stem from southern influence but now recognised as featuring significant indigenous technological and cultural sophistication, in some cases predating cultural and economic expansion initiatives from the south. Porter (2012:1/10) argues cogently for the role of pastoralist groups (often dismissed in traditional models as solely preceding or peripheral to the sedentary 'rulers') as an integral element of the society generating both southern and northern developments.

The early 3rd millennium BC then saw a significant regression in the north, contrasting with the continuing urbanisation in the south. A second phase then emerged in the north, with a '[r]esurgence of urbanism and complexity, c. 2600–2000 BC' (Ur 2010:404) featuring rapid urbanisation with a significant impact on agriculture and animal husbandry. Similarly, a renewed wave of urbanisation took place in the south from the mid-3rd millennium BC, culminating in the late 3rd-millennium BC Ur III period with its highly centralised state economic system, controlling agricultural resources, and eventual urban collapse.

Rethinking working-animal models

There is slowly growing recognition among commentators of the role of the donkey from the late 4th millennium BC, as a valuable complementary technology to the water transport considered central to the operation of southern Mesopotamian trade, distribution and manufacturing, and as an alternative ploughing animal. Contemporary pastoralist groups may have been early users of donkeys for pack; to this day mobile pastoralists employ donkeys for local and long-distance transport including specialisation as caravaneers (Ben-Tor 1986:8; Oren 1989:401), and in many Near Eastern regions today, donkeys are a universal all-purpose adjunct to daily herding of sheep and goats. Anne Porter's (2012) proposal of the key role of pastoralists in Mesopotamian state formation provides an attractive indication of how donkey use might have filtered into an urbanising culture with rapidly growing need for work capacity. Zarins (2014:198) proposes the adoption of the pack donkey as a causal factor in the urbanisation and long-distance trade that is a feature of the Uruk phenomenon, but admits that the chronology points to the donkey as more of a significant supporting factor, taken on late in the 4th millennium BC as a new tool for intensification of existing developments. Brodie (2008:304) suggests a very similar scenario in the Bronze Age Aegean:

> [t]he donkey itself was not an agent of change, nor was it an immediate cause of settlement nucleation or overland trade. It provided a resource that could be drawn upon by communities to provide previously unavailable responses to changing socio-economic and ecological circumstances.

The phenomenon of animal power lay in its utility for fulfilling burgeoning demand, not necessarily in generating it.

Donkeys and the Mesopotamian 'private sector'

In the spheres of both agriculture and urban-led activities such as long-distance trade, under the better-documented state institutions of Sumer/southern Mesopotamia there is growing evidence that a layer of privately operated enterprises existed, from small-scale farming to independent merchants (e.g. Adams 2008:11–14; Liverani 2006:25/35/44/61), rarely recorded by the 3rd-millennium BC cuneiform texts. Asher-Greve (2013:359) points out that '[t]he economic foundation of Mesopotamian polities was agriculture and animal husbandry; most Sumerians and their Semitic neighbours were rural, but nearly all our sources come from cities', focusing on the activities of the urban power elite. Stone (1999:204–12) sees the hierarchical authorities as controlling the provision of arable land, leaving the less accessible or reliable areas for a population delivering crops (and temporary labour) to urban authorities but able to operate smallholdings independently, augmented by exploitation of marsh and desert flora and fauna (Stone 1999:206). Styring *et al.* (2017:9) argue from extensive new analyses of isotopic and other evidence in Syria that

> although our data are consistent with an overall strategy of extensification, this broader framework subsumes a range of behavioural variation that is testimony to a bottom-up, as opposed to top-down, driver of agricultural change. Individual households seem to have practiced a nuanced and flexible strategy in which (1) the crops planted, (2) where they were planted and (3) labour and material inputs of water and/or manure were all fine-tuned to the specific characteristics of the crop, land and/or soil quality, and the highly variable rainfall circumstances of any given year. This diversity in agricultural practice makes sense as a household risk-buffering strategy but not as an elite-controlled share-cropping regime.

Adams' (1981:136–41) seminal report on his survey of the Iraq countryside underlines the apparent scarcity of small rural settlements in the 3rd millennium BC, but Steinkeller (2007) considers that this is a function of their low archaeological footprint, citing texts from late 3rd-millennium BC Umma recording large numbers of hamlets with a grain silo (see an analogy in Figure 2.2). Pournelle reports (2003:84) on the valuable expansion in recent years of Adams' survey initiatives in southern Mesopotamia through new extensive regional surveys, and Cooper (2006:52) on similar work in some regions of the north. Adams agrees (2008:12–14) that in the 3rd millennium BC urban house-plots, date-palm groves and vegetable plots are known to have been privately held and the products exchanged without regulation. In northern Mesopotamia, McMahon (2013:470) suggests that

> [t]here may have been a dual economy: mixed household versus specialised institutional. This variability makes the north's agricultural

Figure 2.2 Ziniaré, Ouagadougou region, Burkina Faso: an African farming hamlet with grain silos. One silo has been opened to make dolo beer

Source: © JG Burkina Faso 2013

> economy appear more flexible and less regulated than Sumer's. However, we have unequal data: botanical remains are poorly preserved in Sumer due to salinisation, and there we rely heavily on texts, with their strong administrative focus, for reconstruction of the agricultural system. Sumer may have had equal flexibility, risk-spreading and crop choice below the administrative radar.

Throughout history there have also been initiatives to conceal assets from the eye of authorities, as for example with enterprising agricultural producers in modern Senegal and Ivory Coast who 'launder' their crop surplus into cattle, which are then kept outside the village to avoid taxation or sequestration (Landais and Lhoste 1990:228).

In a later text, the early 2nd-millennium BC poem *The Debate Between the Hoe and the Plough*, the two agricultural modes are contrasted: the Plough boasts that it is highly regarded by the elite and delivers much greater grain output than the Hoe, but the humble Hoe prevails, as not only does it make furrows as well as the Plough but has multiple other uses (Black *et al.* 1998–2006a). While the Sumerian texts clearly indicate massive centralised agricultural operations, the continuing importance of the hoe and the archaeobotanical record of that period show a more diverse agricultural picture, as would be expected in the presence of smallholdings.

Small-scale farming can feature greater emphasis on use of (female) cows and donkeys rather than higher-cost oxen; Mitchell (2018:3–4) suggests that study of donkey use actively 'provides an entry point in the past of . . . those without agency because they are outside prevailing power structures'. Tibbs

(1989:52–4) reports on donkeys in China as indicators of individual ownership of land: the population of donkeys surged after the 1949 AD redistribution of land to individuals, but then reduced from 1952 as collectivisation (and also mechanisation) took hold, with a new increase from 1978 as new agricultural policies favoured individual ownership of working animals. In Mesopotamia donkeys had begun passing through the countryside on pack journeys as well as being potentially available from mobile pastoral groups interacting with agriculturalists. Female cows were an established presence in the agricultural landscape, a resource-efficient asset for small farmers and, as up to the present day, an option for ploughing.

2.2 Early history of donkeys in the Ancient Near East

Early domestication of donkeys: zones of interaction?

There is extensive specialist literature on the phenomenon, process and dates of early taming and domestication of animals for work; the focus of this book, though, is on the impact of domesticated donkeys (and cattle) as an established presence. There is some value in addressing briefly how and why donkeys were domesticated – and the definition of domestication as applied to donkeys. Does it include *ad hoc* taming and multiple domestication episodes? Does it imply genetic manipulation by selection? This is a particular issue with donkeys, as *ad hoc* practice has been common throughout history and selection has largely been absent. Marshall *et al.* (2014:6153) argue that traditional definitions of domestication reflect 19th-century AD intensive animal breeding practices and that new genetic work is leading some to (p. 6157) 'question whether domestication remains a useful concept' in the case of animals such as donkeys.

More secure clues may lie in the identification of donkeys well outside their natural habitat. Geneticists have determined through (limited) mitochondrial DNA evidence that modern donkeys are descended from two distinct phylogenetic groups stemming from north-eastern Sudan (*Equus africanus asinus* (Nubian) clade) and the Eritrea/Ethiopia/Somalia area (*E. a. somaliensis* clade), with data still emerging on which of the two has had the stronger influence (e.g. Beja-Pereira *et al.* 2004; Kimura *et al.* 2010; Vilà *et al.* 2006; see the Glossary in Appendix A).

It is suggested by some commentators (e.g. Zarins 2014:109) that the ancient range of wild donkeys extended further north and west, to the Libyan desert, and that early *ad hoc* domestication may have occurred here among mobile cattle-herding groups with a need for transportation, as with camp moves, particularly in view of increasing Saharan aridity and the need to reach more distant water sources (Vilà *et al.* 2006:350 among others). More intensive adoption may then have occurred where these mobile groups interacted with sedentary farmers, at the edge of the Nile valley (e.g. Larson and Fuller 2014:126; Zarins 2014:109); this interaction may be illustrated by the

Libyan 'Cities palette' from the late 4th millennium BC, depicting animals including donkeys brought as tribute by desert-dwelling groups from near the western Delta. Meanwhile, domesticated donkeys travelled from Egypt to the southern Levant, on the trail that famously became a vital artery for a time for supplies of wine and oil to Egypt. Uerpmann (1991:12–30) argues that there is also faunal and environmental evidence for the presence of wild donkeys in Arabia in the 4th millennium BC, allowing the possibility of a spread north along rocky zones into the southern Levant; Potts (2011:167) and Zarins (2014:33–6) suggest support for this view.

A promising model for early systematic use of donkeys is therefore the coinciding of mobile groups making low-level use of pack donkeys with intensifying or urbanising sedentary cultures, with new high demand for transportation of farm products, construction materials and trade goods; this resulted in the transfer of a desert species to a cultivated region, with correspondingly greater opportunity of archaeological detection. Demand for donkeys in an area unsuitable for breeding or lacking non-cultivated space is to this day supplied from arid marginal regions.

The road (or boat) to Mesopotamia

While there is still debate on the presence of wild *Equus asinus* in Mesopotamia (e.g. Vila 2014:433), the balance of opinion is that working donkeys in this region were introduced as domesticates. Among the earliest potential indicators of use of donkeys for work in Mesopotamia, and indeed anywhere, are a vivid depiction on a 5th-millennium BC sherd from Fars in highland south-west Iran of a donkey with a possible blanket or pannier (Figure 2.3) and faunal remains from 4th-millennium BC Tell Rubeidheh on the route between the Zagros mountains and southern Mesopotamia (Payne 1988). The presence of domesticated donkeys in Mesopotamia from the late 4th millennium BC is then suggested by still fragmentary faunal and other evidence (e.g. Zarins 2014:67/70/78/188/250). Kathryn Kelley (pers. comm. 2017) reports on the proto-cuneiform sign KUR used in unprovenanced late 4th-millennium BC Archaic Texts in relation to both domesticated male donkeys (but not other adult domestic animals) and human workers, in Uruk and Jemdet Nasr texts, with connotations of both being brought in from elsewhere – possibly from the mountainous areas to the east and north – for work. Both are effectively categorised as domestic servants, under the term ḪAR.TU; at this stage there is no specific word for animal, as distinct from human, in Sumerian text.

There is some evidence that donkeys reached Arabia at an early stage, and a plausible suggestion is that they were brought across the Arabian Gulf by coastal boat (Hans-Peter Uerpmann pers. comm. 2015), specifically for use as a valuable means of pack transportation up into the highlands and then over the Zagros range into Mesopotamia. Donkeys may also have arrived from the west, ferried across the Mesopotamian rivers. Zarins (2014:199)

Figure 2.3 5th-millennium BC sherd from Tol-e Nurabad in Western Fars, Iran
Source: Potts 2011:170, by permission

reports an Old Akkadian 3rd-millennium BC text concerning donkeys and bovids transported by boat, while donkeys have been taken on rafts down the Mesopotamian rivers from early times. In more recent periods, European and other settlers took donkeys to southern Africa on their ships.

In the 3rd millennium BC, faunal, textual and representational evidence of donkey (and hybrid) use is more common (e.g. Figure 2.4). Postgate (1986:200) comments

> I had not appreciated the fact that the presence of the donkey in 3rd millennium Mesopotamia, which is, I think, taken for granted by most cuneiformists, would be a matter of surprise for zoologists.

An early 3rd-millennium BC Sumerian text describes how 'the people of Aratta stepped forward to admire the pack-asses' sent east over the mountains with grain to exchange for precious goods (Black *et al.* 1998–2006b

Figure 2.4 Impression of a cylinder seal, ED III (c.2400 BC). Upper register: equids (donkeys or hybrids) pulling a plough with a seed-funnel

Source: IM 83755. Photo courtesy of the late Dr. Lamia Al-Gailani

Enmerkar and the Lord of Aratta; though Liverani (2006:44) considers that while donkeys were present, this was an ideological myth and that grain was not a common long-distance commodity). Pre-Sargonic (mid-3rd millennium BC) depictions and texts (e.g. in Maekawa 1979:102/111), and texts from late 3rd-millennium BC Umma and Girsu relating to fodder cost, inspection of plough animals and land assignment (Heimpel 1994:18/22), give evidence of the established use of donkeys for ploughing. Heimpel (1995:134) calculates that the ratio of donkey to cattle teams in Lagash and Umma was respectively 3:4 and 1:3; the texts indicate that ploughing donkey numbers on state-controlled farms flourished in Ur III though were dwindling by the beginning of the 2nd millennium BC. Zarins (2014:193–5) gives tables too of the use of donkeys versus cattle in state-controlled agriculture at Girsu in the Ur III period; and more generally his comprehensive 2014 work, *The Domestication of Equidae in Third-Millennium BCE Mesopotamia*, forms an invaluable companion source-book on the zooarchaeological, textual and representational evidence to date of working donkeys in the 3rd millennium BC (and earlier) in Mesopotamia and also in the southern Levant and Egypt. In re-examining many of the classic equid finds, he has in particular recorded significant examples of severely inaccurate identifications and faulty contextual information (Zarins 2014:53–65).

Meanwhile, as the presence of donkeys spread outwards through the regions, it could be envisaged that they became adopted increasingly in various roles in small-scale farming, under the radar of state operations and so of textual records. This invisibility persists for a similar reason today, with use of donkeys on small farms worldwide often excluded from official survey data (Kreike 2010:100 and others) despite their huge numbers in developing regions.

Elite burials and donkey status dichotomy

There are suggestions that in Egypt donkeys might have been first captured for ritual or ceremonial use, and certainly transport animals in the Ancient Near East are frequently linked to ceremonial or sacrificial events, from the numerous apparent ceremonial draught-animal burials in the 3rd millennium BC, alone or in elite human graves, to the 2nd-millennium BC Mari texts mentioning donkey-killing to seal a treaty. In later 3rd- early 2nd-millennium BC texts the donkey (or very possibly a hybrid) is referred to as a noble and spirited animal, as in the poem in praise of the king Šulgi (Black *et al.* 1998–2006c:Šulgi A ¶16–18), who is famed as a fast runner and where he is depicted as

> I am a mule[i], most suitable for the road.
> I am a horse, whose tail waves on the highway.
> I am a donkey of Cakkan, who loves running.
>
> [i] *Or 'Nisku' may be translated as 'choice donkey' (Zarins 2014:175)*

Zarins (2014:53–65) advises caution about some of the traditional references to the phenomenon of equid burials with high-status humans: having re-examined some of the studies and actual faunal evidence, for example the elite burials at Ur and Kish, he concludes that some of the equids were either not associated with the human burial or were entirely misidentified. On closer examination, too, equids in elite texts, representations and burials are commonly the expensive and prized onager-donkey hybrids rather than donkeys (Postgate 1986:194–200; Weber 2008).

Separately, several commentators (notably Milevski 2011:192/292) suggest that special treatment of donkeys is, rather than ceremonial, the result of the impressive spectacle of large, fast, strong animals submitting to humans and performing valuable transport and communication work for the greater wealth and glory of their merchant or noble owner. Greenfield *et al.* (2018:202) point out that donkey-burials in the 3rd millennium BC, in contrast to those in the 2nd millennium BC, are often not associated with temples or monumental structures. A conundrum throughout history is the positioning of the donkey as simultaneously elite and despised; Milevski's (2011:233) hypothesis of elite human burials with donkeys as relating to merchants, further developed by Greenfield *et al.* (2018:202) and Shai *et al.* (2016:6–7), may help to resolve this. Donkeys were a valuable resource to merchants benefiting economically and socially from their ability to move goods long-distance at low cost, and so while lowly and despised in themselves a symbolic pair from the 'fleet' of pack donkeys would have been appropriate adjuncts to the burial of one whose power derives from control of this facility (as perhaps with slaves interred with their masters), whether manifested as actual animals or as figurines.

> EB II and III urban centers with their large populations could have had their own group of merchants as in the case of Ebla palatial economy of the third millennium BC (Pettinato 1979) or that of second millennium merchants in Ugarit (Rainey 1963). Such likely analogies suggest that the existence of a cult related to donkeys as beasts of burden was related to the existence of a group socially differentiated from the rest of the population by its economic activities, i.e. merchants and/or donkey herders. (Milevski 2011:232)

The social differentiation indicated by donkeys in burials may therefore be horizontal rather than vertical: a specialist rather than higher-class caste, with the prestige element applying to the donkey employer not the donkey itself. Milevski (2011:177/196/232–3) and Greenfield *et al.* (2018:202) suggest that these burials and figurines might also relate to donkey breeders, traders or drivers, so a cultural marker (Silvia Prell pers. comm. 2017) rather than a signifier of high status. These groups were also likely to have been socially distinct, and if more recent parallels can be employed, these may have consisted more of low-status castes whose close association indeed lends denigratory connotations to donkeys, who are in themselves perceived as difficult by those inexperienced in their use. Cochin (1995:62) records that in West Africa an untouchable caste breeds donkeys and suggests that strongly hierarchical societies may classify their domestic animals into hierarchies, assigning donkeys to women and children and to low-caste groups.

Today, as in antiquity, practical attitudes to ownership of a useful asset go hand in hand with an extremely long tradition of pejorative attitudes to donkeys: the term 'donkey' was often an insult in ancient Egypt and Mesopotamia as now (see notably Power 2004:131–4 on Egypt). In Egypt the donkey became associated with the god Set and with notions of evil; Osborn and Osbornová (1998:135) suggest that 'the donkey eventually became the 'supreme scapegoat' and that its image was destroyed in magic rituals'. Power (2004:131) notes that

> [t]he donkey is perhaps one of the most important, albeit neglected and misrepresented animals in history . . . this animal's relentless work and almost inexplicable endurance over possibly 6000 years of domestication is overshadowed by one of the worst reputations in the animal world . . . for being stubborn, unruly, and hard to control – a reputation endorsed by many of the secondary sources in the handful of references to donkeys that exist within the Egyptological literature.

She suggests (pp. 132–4) that donkeys are ascribed low status because of their loud brays and their habit of frequent copulating and erection in public (Figure 2.5), recording their common depiction 'with spontaneous erections and masturbating' (p. 133). Marshall and Weissbrod (2009:72) suggest a very similar explanation in modern Kenya for 'the apparent and pervasive disjunction between the importance of Maasai donkeys for survival and their relatively low status':

> because of their slow growth rate donkey herds are not ideal for accumulation of wealth, and donkeys are not considered to be food. Donkeys in Africa are also well known for the length of the penises, their sexual activity and, when they are not herded, for copulating in public places. Furthermore donkeys are noisy and their bray is intrusive'.

A note on onager-donkey hybrids, and horses

Equus hemionus (which I refer to in a Mesopotamian context as the onager) is similar in appearance to the wild donkey (*E. asinus*) and native to the arid belt from Syria to Mongolia, and perhaps in the southern Levant, with its range and population size much reduced since antiquity. There is debate as to the genetic relationship in the far past between *E. hemionus* and *E. asinus* (Geigl and Grange 2012:92; and see the Glossary in Appendix A), but by the period under study the separation is clear. The distribution of onagers possibly overlapped with the distribution of wild donkeys, though the two were

Figure 2.5 Misbehaving donkeys in 3rd-millennium BC Egypt. 'Figure 1. Tomb of Neferirtenef [17]'

Source: Redrawn from Power 2004:144, by permission

divided by their differing habitat preferences, with onagers preferring flat sandy regions and donkeys rocky hills (Groves 1974:110; Uerpmann 1991).

There is evidence from the 8th/7th millennium BC onwards of the hunting of onagers for meat and perhaps hides in Mesopotamia and the southern Levant; they were hunted in the 4th and 3rd millennia BC and later (e.g. Vila 2006:103–8/113), and even up to modern times. The onager's intractable and restless temperament contrasts with that of the donkey, and it is now increasingly agreed, from faunal and textual evidence, that although onagers may be tameable as foals or temporarily held for hybridising, they were never domesticated for work; Zarins (2014:14–32/45–7/65–7) makes a comprehensive case for some interpretations of historical accounts and earlier texts on domesticated equids falling prey to confusion with donkeys and hybrids. Littauer and Crouwel (1979:26–8) describe the difficulties of determining equid type in 3rd-millennium BC representations, with (even assuming that the artist is drawing from life) ear length and mane and tail style the most indicative.

Hybridisation of the onager with the domesticated donkey is likely to have been pursued for the sake of the heterosis (hybrid vigour – good size, hardiness and ability to survive on poor food) common in crosses and demonstrated in the horse-donkey cross mule; use of native onagers also extended the stock of imported or rarely-obtained donkeys despite the disadvantage of the sterility of the offspring. Texts and representations are, so far, the main indicators of the emergence of onager-donkey hybrids in the early or mid 3rd millennium BC, but a new era has been heralded by the identification by mitochondrial DNA of the equids ceremonially buried at Umm el-Marra at the end of the 3rd millennium BC as being male onager/female donkey crosses and showing signs of use for work (Weber 2008:501–5). At 3rd-millennium BC Tell Brak (ancient Nagar in NE Syria), seal impressions indicating provision of hybrids to ceremonial centres are closely associated with evidence of the careful keeping of equids and of valuable objects (Oates 2001:279; Oates and Oates 2001:41–7); these indications are reinforced by accounts in 3rd-millennium BC tablets from Ebla of the flourishing hybrid-breeding centre at Nagar (Archi 1998:10) and of Ebla's additional approaches to centres further east with requests for good hybrids. The Ebla and Diyala texts record the prices for hybrids rising sharply in the 3rd millennium BC, to 6 or even 40 times the price of a donkey (Zarins 1978:14).

By the 2nd millennium BC the onager was replaced in Mesopotamia by the horse for hybridisation use (e.g. Zeder 1986:406); however, the horse-donkey hybrid (the mule/hinny) is excluded from the scope of this book, as modern usage and management comparisons with the onager-donkey hybrids of the earlier period upon which I am focusing would be inappropriate given the genetic influence of the very different physical and behavioural characteristics of the horse. Onager-donkey hybrids are still produced in India and Iran today, but accounts of their usage are too limited for present purposes.

Horses are native to the Central Asian steppes and were probably first domesticated there (perhaps initially for food) in the 4th or possibly 5th millennium BC; wild horses may also have been present in Anatolia and Iran, with the domestication concept filtering into these regions, or domesticated horses may have been imported and passed on into Mesopotamia and the southern Levant. The earliest textual and iconographic evidence of horses appears during the 3rd millennium BC, but horses are apparently rare until the early 2nd millennium BC (Postgate 1986:195).

2.3 Detecting working animals in the archaeological record

Zooarchaeological detection of working donkeys

The detection and broader acknowledgement of the presence of working donkeys in the Ancient Near East are both considerably impeded by the low incidence of donkey skeletal remains: in many regions throughout history and to the present day, there is marked evidence of cultural (and practical) reluctance to eat working equids, with corresponding scarcity of bones. It is sometimes insufficiently recognised that the presence of animals which are used for work but not generally eaten – so not appearing in food-middens – can be seriously underestimated (Goulder 2018:81). The important issue of the non-eating of donkeys is a large and sometimes inexplicable one, and I have addressed it in some detail in Chapter 4.2.

Mesopotamian archaeology, in its long history during the 20th century AD, suffered a measure of unavoidable bias and incompleteness, and was initially influenced by cultural preoccupations of the time in its excavational emphasis on monumental architecture and elite material culture. It also reflected the means of site identification available at the time, with its inevitable focus on tells – the occupation mounds that dot the region. This emphasis is certainly changing, but the archaeological legacy still results in most animal-bone finds coming from food-refuse middens, often in urban environments. Such middens form a key source of evidence of working cattle, as the latter are almost universally used for meat at the end of their working lives; but there is still a paucity of recognition of the markedly different deposition processes that apply in the case of donkeys, with distribution variations perhaps more indicative of differences in cultural mores on the eating of donkeys. (The phenomenon of 'elite' burials is a separate issue)

The remains of donkeys are unlikely to be found where they work, as for practical reasons (from modern analogy) they will be removed and their bones perhaps destroyed and scattered by predators and scavengers, resulting in a find-pattern that cannot be taken to reflect the living animals' incidence or distribution. Where remains are found, key skeletal elements may be missing, and in the general circumstances of limited comparison material for domesticated donkeys, interpretation of pathologies is also a concern.

Zarins (2014:82) sums up the resulting conundrum for the detection of working donkeys:

> [g]iven such a purpose for this equid, its remains would not turn up in house or public building refuse dumps and debris unless fortuitous circumstances happened to occur, e.g. in the case of the Abu Salabikh dump donkey. . . . [I]n a large-scale excavation within an urban area, unless the excavator deliberately was looking for stables, stalls, burials, etc., the remains of the pack donkeys . . . would not turn up.

The issue of Equus hemionus in Mesopotamia

A significant additional obstacle in the study of early donkey use in Mesopotamia is the difficulty of distinguishing *Equus asinus* skeletal remains osteologically from those of onagers (*E. hemionus*) (Milevski 2011:178 and others) without costly DNA analysis. The presence of their remains can impede detection of domesticated donkeys: site reports from the region typically list 'equid' remains, with a qualitative discussion of any evidence pointing to *E. asinus* or *E. hemionus* identity. To give an example, at 3rd-millennium BC Tell Sweyhat in northern Mesopotamia Jill Weber (1997:137) (commenting that '[e]quids are notoriously difficult to distinguish according to species, especially when ass and half-ass [onager] are in question') concludes that among the twenty equid bones recovered, three or four may be from (domesticated) donkeys.

This would be less of an issue in Mesopotamia if domesticated donkey remains exhibited clear evidence of selective breeding that might assist in telling them from onagers; but unlike most livestock species, domesticated donkeys are not easily distinguishable morphologically from wild ones. The classic Near Eastern illustration is of the well-known donkey burials at Abydos in Egypt, c.3000BC, where the animals have work pathologies but are otherwise similar to contemporary wild donkeys. From earliest to modern times, donkey-using cultures have encouraged free breeding with the wild and made little use of selective breeding; unlike species kept for meat or milk, the size and vigour of wild donkeys are prized virtues. I discuss in Chapter 4.1 the persisting tendency among some zooarchaeologists to detect donkey domestication by size reduction, despite strong modern and ancient evidence for the limited validity of this measure.

Work pathologies are of course a valuable indicator – exostoses, joint damage, osteoarthritis, particular muscle-attachment development, traumas; these have been well-discussed by zooarchaeologists for cattle in particular. Work on detection of skeletal alterations from employment for pack work – differing from traction traumas – has been more limited to date (Russell 2012:228). While use for pack may be generally visible in the vertebrae, Shai *et al.* (2016:11) observe for example that donkeys with their narrow backs are typically given paired side-loads, with different resultant pathologies from broad-backed pack animals, and from traumas caused by riding. Shai *et al.* (2016:11) also comment on 'the difficulty of separating out natural variability and congenital defects from actual pathologies', and there is a risk here of equifinality.

Some other zooarchaeological indicators are now emerging. Aside from work pathologies, a range of osteological and particularly dental indications of constraints upon natural moving and eating may be detectable (Shackelford *et al.* 2013:4170–71; Shai *et al.* 2016:11; Zarins 2014:42): evidence of fodder-eating from teeth wear and isotopic bone analysis, crib-biting (though equids also gnaw tree-branches in the wild), rickets indicating indoor living, deciduous teeth found in enclosures, and trauma from hobbles. The issues of lip/nose rings and bit-wear on donkeys of this period are debated by specialists and I shall not enter into them here; donkeys are rarely bitted for pack or traction. Finally, stable isotope analysis could play a role at times in identifying equids originating from far away, indicating perhaps donkeys bred elsewhere.

As with cattle, a preponderance of aged and disabled/injured donkeys in the record (except in ritual circumstances) might indicate working use, as there are practical arguments against killing young, fit working animals. This model, though, again does not reflect the unique position of working donkeys: aged donkeys no longer fit for work are commonly left outside settlements to die or be killed, but are also sometimes eaten, with a confusing resultant deposition pattern. On a related subject, Grigson (2003:220) and Ovadia (1992:19) consider that as with cattle (see what follows), there is a possible case for inferring domesticated donkeys from the percentage of equid bones in the total animal record increasing over time at a site. But aside from the significant disposal bias and issue of confusion with onagers noted previously, such assertions would need to be supported by further arguments for their not being used (solely) for food or hides.

Issues with detection of working cattle

The cattle 'percentage increase' model

One of the means proposed for detecting the advent of ox-ploughing at Ancient Near Eastern sites is calculation from faunal remains of the percentage of cattle in the total animal record in the settlement: an increase to 20% is suggested as the threshold for use of deep ploughing with oxen, for example in the Late Chalcolithic at sites in the southern Levant (Wapnish and Hesse 1991:26), compared with Sasson's (2010:56) suggestion of a common proportion of 15% at Near Eastern sites. Sherratt (1981:289) assumes that a short ploughing season would require the keeping of a herd of cattle, and some modern analyses of working cattle management also conclude that a 'supporting' herd of some size may be necessary for the secure provision of the needed number of working animals at any one time. McCann (1995:70), for example, points out that use of oxen for ploughing brings with it a need for a livestock system which produces, trains, and maintains a stable supply of oxen; for a pair of working oxen this might require (McCann 1984:6, 1995:48, referring to modern Ethiopia) ten head of cattle (two oxen, two young 'apprentice' bulls, a stud bull, and four or five cows). Peters (1986:138) reports an estimate that a herd of 30 cattle is required to maintain four draught oxen, and Lawrence and Pearson (2002:101) calculate that to maintain 100 adult working cattle, a total herd size of 130–350 is required, depending on birth rate, the age at which animals start work, the age at which they die and which sex is used for work.

Other commentators differ on the validity of this model. Sasson (2010:48) examines both recent usage of working cattle in Israel and evidence of their presence at sites there in antiquity and concludes that 'cattle abundance does not point to agricultural intensification'. The 'threshold' assessment, too, does not take account of growth in non-plough uses of working oxen, such as transport and other agricultural traction. The calculations assume that only male cattle are used for work, which is by no means the case in many regions worldwide today. These calculations, too, do not take account of the incapability of small-scale farmers of maintaining a 'supporting' herd, particularly in regions where cattle do not thrive, with instead a common informal system of a community-wide reserve of oxen, often spread around the area, perhaps in grazing herds many kilometres away, with a sufficient number of ready animals emerging as needed and exchanged and shared among the wider community (e.g. Farnham 1997:105). Experienced users, too, may recruit cattle *ad hoc* for work from their general herd, making the presence of working cattle hard to detect archaeologically in what can seem largely a traditional meat/dairy herd. New cattle for work may also come from specialist breeders far away, as occurs to this day.

Finally, account is not taken in the cattle-threshold model of cattle herds kept for prestige reasons. Russell (1998:49) lists likely archaeological indications of cattle kept for prestige or wealth: reduced animal size, cattle kept in areas 'not optimal in terms of subsistence' and far from settlements, indications that individual households are trading up to cattle from small stock, and cattle featuring in ritual and symbolism. In parts of modern Africa, cattle herds are often still maintained largely for status reasons, with their care sometimes subcontracted to pastoralists.

Cull profiles and castration

The caveats mentioned apply similarly to models based on cull profiles, as summarised by Johannsen (2011:13):

> [i]t has sometimes been assumed that draught exploitation will manifest itself in faunal assemblages as a relatively high proportion of bones

> from old (castrated) males. However, this assumption is not necessarily correct in all contexts. Historical sources, ethnographic observations and literature on modern agro-economic and farming development reveal substantial variation in known draught exploitation strategies. This variation highlights to zooarchaeology the importance of supplementing mortality patterns with other lines of evidence, such as artefacts, iconography and osteomorphology.

Gaastra *et al.* (2018:1464) concur that short-term *ad hoc* use of animals for work 'would not necessarily alter management profiles or significantly affect bone development across an entire faunal sample'.

In Chapter 5.1 I comment briefly on the value and limitations of osteological evidence of castration. While presence and age of castration is a potential indicator of a working ox, bulls are also castrated for other reasons (and not always castrated for work), and it does not detect working cows. Meanwhile, donkeys are not commonly castrated.

Representations

While depictions of working animals in antiquity are not strictly zooarchaeological evidence, they are studied by zooarchaeologists and so may have a place here. I refer throughout this book to the limited relevant text and (to a lesser extent) representational examples relating to use of working animals in antiquity, in examination of parallels with modern practices. Zarins (2014:148) argues that 'it is likely that three-dimensional representations, figurines sculpted by those familiar with livestock and who know first-hand patterns of domestication, will afford the most reliable indicators of zoological data'. However, the inevitable bias in these depictions towards elite concerns and knowledge – and perhaps to traditional rather than current practice – contributes considerably to the invisibility of donkeys in this period. Pournelle and Algaze (2014:21) suggest that Uruk-period iconography forms part of an idealised Soviet-Russia style depiction of farming processes under the aegis of controlling authorities, perhaps to the common exclusion of donkeys:

> Irene Winter (2007) has recently reviewed the pertinent art-historical evidence and notes the incessant visual reference that the art of the time makes to domesticated grain and livestock, which she interprets as a self-conscious attempt by urban elites to legitimate themselves as providers of unending agrarian resources.

3 Donkeys versus cattle

3.1 Donkey and cattle behaviour and physiology

Introduction

Here, as a prologue to my chapters on the use of working donkeys and (to a lesser extent) cattle, I set out how the differing behavioural and physical characteristics, capabilities and potential of the two species have played a crucial part in their likely work trajectories in Mesopotamia and elsewhere. The sections that follow, on the comparative physiological and behavioural irreducibles of donkeys compared with cattle when used for ploughing, and more broadly for work, aim to provide a practical briefing as a background to this process in antiquity.

The behavioural patterns of donkeys differ significantly from those of cattle. Donkeys live naturally in small sociable groups, so adapt well to working alongside humans; they are quick to learn and require little supervision. The physiological differences between donkeys and cattle have also led to complementary roles. Donkeys are strong for their feed input, low-maintenance and have a longer working life. Cattle have greater absolute power (see Table 3.1), if this is needed for ploughing heavy soils, but a higher purchase cost (and the issue of theft due to their greater value) and much more intensive feeding and watering needs.

There has been significant archaeological interest in the impact of climate and environment on various innovations in antiquity, including choice of working animal. Modern development studies are sometimes inappropriate for research on this, as the choice of working animals initially introduced into sub-Saharan Africa in the 20th century AD in particular has commonly related more to centralised official strategies than to physiological or behavioural suitability. But modern studies demonstrate strongly that, in practice, in the majority of environments in developing regions, both donkeys and bovids perform satisfactorily for ploughing (Figures 3.1 and 3.2). It is largely in extreme environments that physiological differences come to the foreground in choice of ploughing animal; so in regions with heavy soil, larger animals such as oxen are commonly preferred, while in hot, arid regions with light soil and limited grazing and water, the desert-adapted donkey may

Figure 3.1 Donkeys ploughing in western Ethiopia. Note the unsuitable yoke for the donkeys.

Source: JG Ethiopia 2014

Figure 3.2 Cattle ploughing in the same field. A farmer told me that the donkey team ploughed around stones while the cattle team were able to push them aside

Source: JG Ethiopia 2014

be favoured. Archaeological consideration of working cattle and donkey choices in Mesopotamia could be said to have focused overly on suitability for ploughing, whereas outside extreme environments other factors may well prevail: other employment of the animal (notably local transport, and meat and milk), comparative maintenance costs, availability/familiarity and cultural preferences.

Behaviour

Donkey behaviour patterns are often misunderstood as they are thought of as 'small horses' (McGreevy 2004:30). This is misleading as the two species have major behavioural differences in the wild, arising from their different habitats, which are reflected in their differing responses to situations in a domesticated environment. Horses originate in the cool grassy steppe and donkeys in hot, arid North Africa. In arid regions, where food and water are scarce, a loose social structure allows individuals to roam long distances daily, keeping the density of animals down while allowing some social linking for protection; donkeys' loud bray, audible over several kilometres, enables them to remain in touch across sparsely populated areas (McGreevy 2004:162). Donkeys are described by some as herd animals, but a better term is sociable: in the wild, they form shifting associations with a small number of other individuals (e.g. McGreevy 2004:139; Marshall and Weissbrod 2011:S404); therefore while they actively enjoy company, of their own or another species, and readily team up with other donkeys or humans, unlike oxen they also work well alone.

> Almost all working animals are of species that naturally form groups, and their relationship with humans seems to involve regarding humans as members of their group, usually in the capacity of leader, or at any rate guardian. (Jones 2008:5)

In small groups, as opposed to herds, signs of stress or pain are seen as indicative of dangerous weakness and the animal is excluded from the group, or if defending a territory is targeted for attack; in contrast in herds, reactions to threats alert the whole group so are seen as advantageous. The small-group characteristic translates into the well-known patience and stoicism of donkeys, as they are behaviourally adapted to showing few signs of pain (e.g. McGreevy 2004:30).

Unlike full herd animals, which have a strategy of fleeing from a threat (as predators should only catch the hindmost), donkeys become immobile or group and face a predator, as a good strategy in a small group under threat. They have a natural highly developed sense of individual self-preservation, and their strategy is to 'freeze' and assess situations and obstacles cautiously before making a move, whether in dealing with a predator or if they do not understand what they are being asked to do or why (e.g. Yilmaz 2012:21/33).

This can be misunderstood in working situations as uncooperativeness or stupidity, and in the Western world and elsewhere donkeys are famously considered to be stubborn and difficult. Power (2004:131–6) also reports the dichotomy of view in 3rd-millennium BC Egypt, where donkeys are described in the (urban) texts as difficult while the depictions in tombs commonly show them as cooperative (Figure 3.3).

In studies of modern use in Africa and Asia, donkeys are widely agreed to be less demanding than oxen to control, to learn quickly from other donkeys and from humans, to remember their training longer than cattle and to carry out tasks with minimal or no supervision (e.g. Sosovele 2004:109; Yilmaz 2012:57). They are not generally aggressive to humans and can be handled, harnessed and worked by a single individual, including women and children. The rare early 3rd-millennium BC depictions of equids occasionally show them muzzled (as in the Standard of Ur), perhaps as they were onager-donkey hybrids retaining some of the latter's wild behaviour. In modern times donkeys are very rarely muzzled unless working among growing crops, though Littauer and Crouwel (1979:30) suggest an additional element, in the case of ungelded males, of restraint from 'bickering among themselves'.

Donkeys have a good memory for routes: in the wild they range long distances in search of preferred food and can find their way back to their home territory (e.g. Förster *et al.* 2013:195), and owners often make use of domesticated donkeys' ability to return home alone. I heard from donkey owners at the Yehil Berenda grain market in Addis Ababa (JG Ethiopia 2014) how they led 'taxi' donkeys laden with goods to a customer's home and then dismissed the donkey to return to base unaccompanied. Starkey (1995:148) too reports on donkeys sent unaccompanied daily down the hill from a village in Latin America where they are loaded with water and return

Figure 3.3 Cooperative donkeys in 3rd-millennium BC Egypt. 'Figure 16. The tomb of Werirenptah [6]'

Source: Redrawn from Power 2004:147, by permission

solo, and there are accounts in Africa and elsewhere of donkeys regularly sent unaccompanied across national borders laden with smuggled goods, an employment hardly possible for cattle.

This and their propensity to follow a leader allows large groups of donkeys to be controlled by one person on regular routes. Donkeys in the wild naturally follow a leader in single file, so easily learn to walk in straight lines and recognise furrows, while oxen are perhaps blunter instruments (see what follows on their physiology).

Separately, donkeys' instinct to stand and face predators has led to them gaining popularity as guard animals for cattle, sheep and goats in the USA, Canada, Australia, South Africa and elsewhere, and it is estimated that 25,000 guard-donkeys exist worldwide. A guard-donkey is trained by accustoming it to the flock, and can guard up to 400 sheep, goats or cattle; this role combines donkeys' sociable propensities with their aggressiveness towards predators: with their sharp hooves and strong teeth, donkeys attack and sometimes kill coyotes, jackals, wild dogs and other predators (e.g. Yilmaz 2012:23).

Behaviourally overall, donkeys offer significantly more potential (and actuality) than cattle in replacing a range of working humans. Cattle can be trained to respond to aids and commands, but largely offer extra force. Donkeys work unaccompanied – guarding flocks, transferring goods long-distance and water-containers locally – or at most guided by a young child, freeing adults for other work or leisure.

Physiology

Oxen have powerful shoulders but a relatively weak chest, and a yoke is designed to be powered by the shoulders, with the yoke held forward by the ox's strong withers. Donkeys have low, bony withers, with an equid's long slender thin-skinned neck and muscular concentration in the chest; early depictions of state-controlled animal ploughing in Mesopotamia indicate that it was normally carried out with a pair of animals and a yoke, which militated against donkey use as the yoke is not suited to their body shape and restricts the power that they can deliver.

Paired oxen are hard to turn as they are bulky and have relatively inflexible necks, as well as having insensitive hides and so may not move their head away from the touch of a whip; ploughing oxen generally therefore need to be led round turns, and the turning circles need to be large. Makwanda *et al.* (2000:175) describe how in Tanzania the weeding pattern for a pair of oxen involves moving across the field and back in alternate blocks of five rows, thus avoiding sharp turns at the headlands. In contrast, donkeys have a flexible neck designed for browsing and for watching for dangers while feeding; this, and their narrow bodies, allows them to turn sharply. Hagmann and Prasad (1995:235) add that oxen, being bigger, damage crops more than donkeys during weeding, and unlike donkeys do not naturally tend to walk in straight lines; so donkeys, if available, are often preferred for pulling precision instruments such as seeders and row-planters.

Ploughing practices, and the animals used, can provide detectable field configurations; square-ish fields are preferred for cross-ploughing, but long runs for single furrowing and even S-shaped fields to facilitate turning of large, cumbersome teams and ploughs, as developed in Mediaeval Europe (Brigden 1984:8). In the 4th and 3rd millennia BC in southern Mesopotamia, centrally organised irrigation systems resulted in long, narrow fields – well-suited to ploughing with animal teams – between furrows channelling water from canals (e.g. Liverani 2006:16–17). Heimpel (1995:93) has analysed late 3rd-millennium BC texts from Girsu with the aim of correlating use of ploughing donkeys with the number of furrows seeded in different areas, though with no specific conclusions.

Donkeys in the wild are capable of running at 40–50 km/hour, for example during breeding behaviour or dealing with predators (Shackelford *et al.* 2013:4172–7), and have a lower-leg conformation which resists shock when moving at speed; they have much better acceleration than cattle, though they have a jolting gait at speed, reducing their desirability as a riding animal. Donkeys have a uniquely efficient ankle construction which allows very high energy transfer in motion, well beyond that of cattle and humans, and a gait that minimises energy-consuming vertical oscillation, especially in rough terrain (Yousef *et al.* 1972:337–9). Dijkman (1992:155) also suggests that 'non-muscular structural elements' in the donkey's anatomy (a feature also noted in African women head-loading) may contribute to its notable ability to carry heavy loads. The donkey's energy cost of standing is somewhat higher than cattle, but the energy cost of walking is only half that for cattle, and that for pulling also less (summarised in Jones 2008:12; Tisserand and Pearson 2003:64).

These energy-saving adaptations, which enhanced survival in desert conditions, increase the donkey's suitability as a work animal in hot climates with limited food and water supply. Donkeys are widely evidenced as being able to go without water for up to three days; they can withstand up to 30% body-weight loss by dehydration and can then rapidly take up a full day's-worth of water (Yilmaz 2012:71). Dehydration at high temperatures reduces donkeys' food intake but appears to increase digestibility of ingested dry matter (Förster *et al.* 2013:195). I discuss in detail the different grazing and foddering needs of cattle and donkeys, and their effect on comparative performance, in Chapter 5.2. Descriptions of the utilisation of donkeys in the present and past, particularly for pack in desert environments, underline their hardiness and ability to endure fatigue and difficult conditions over long periods, well beyond those of other working animals until the advent of the camel.

Cattle can be unstable on slopes and stony areas but come into their own on heavy, boggy ground (Lindblom 1931:32), while donkeys are at home on rough, rocky ground and have good balance on slopes. The position of donkeys' eyes allows them to see all four feet at once, making them very footsure (Yilmaz 2012:23). Donkeys have narrow, hard hooves suited to their native arid, stony desert conditions, in contrast for example to those of horses from the grassy steppe; donkeys are known to be able to dig for water.

Donkey working life

Data on donkey reproductive speed, age of starting work and particularly lifespan are often based on accumulation of experience, with little meaningful statistical material; the notes in this paragraph are a synthesis of numerous accounts from countries making significant use of donkeys. Donkeys are much longer-lived than cattle, often living beyond 30 years in suitable environments; they are disease- and parasite-tolerant and show few clinical symptoms and signs of pain. In difficult working environments such as Ethiopia life expectancy – or at least working life – may be 13 years or less, with a generally longer working life on farms than when employed by pack transporters (Pearson *et al.* 2001:43); in very humid areas they may only live for a couple of years from their arrival from the arid zones where they are bred, dying of trypanosomiasis or other tropical-region diseases (JG Ethiopia 2014). Donkeys can carry loads from the age of 2½–3 years, and their longer potential working life than cattle makes for a better lifetime return on breeding and training, though the higher breeding rate and meat value of cattle can be set against this.

Maekawa reports (1979:111) that donkeys in mid-3rd millennium BC Mesopotamia were used for traction from the age of two years; Zarins (2014:256) reports more broadly from analysis of 3rd-millennium BC Mesopotamian texts that donkeys at the time seem to have started work at a similar age to now, but to have reproduced more slowly and died younger.

Zebu versus taurine cattle

It should be borne in mind that modern African data on cattle physiology and behaviour often relates to animals with at least some zebu blood, and care is necessary when drawing inferences about cattle performance in the Ancient Near East before the later advent of the zebu strain. Early domesticated cattle in Africa were taurines, as in the Ancient Near East and Europe, but from the 1st millennium AD these were increasingly hybridised with zebu (*Bos indicus*) stock from Asia via Arabia. Zebu have various physiological advantages over taurines in hot, dry environments, and better resistance to many diseases, and are favoured as they are larger than African taurines (e.g. N'dama); however, the latter are fully heat-adapted and have more resistance to tsetse-fly-borne trypanosomiasis and so are preferred in some African regions.

Until the 4th millennium BC only *Bos taurus* was known in Mesopotamia, though this does not rule out comparison with sub-Saharan African cattle: Galvin (1987:123–6) notes that 100% *Bos taurus* animals in modern Syria are fully heat-adapted, with convergence of characteristics with the zebu, and that 'genetic changes reflecting product specialization in bovids . . . had taken place by at least 3000BC' (Galvin 1987:123) in Mesopotamia. Meanwhile, *Bos indicus* may have arrived at the eastern margins in the early 3rd millennium BC, and in southern Mesopotamia by the mid 3rd millennium BC, though, so far, the evidence is from representations rather than faunal evidence.

Energy and work rates

A theme of this book is that assessment of the relative benefits of using donkeys versus cattle in a farm environment is vital but extremely complex. Many published calculations, for example of hectares cultivated per day, ignore a wide range of social and practical factors; in practice, as is demonstrated daily in ancient and modern farm-based circumstances, farmers make rule-of-thumb decisions about the most suitable animal type and system to employ.

Calculating the comparative practical performance of working animals includes assessment of speed, force/output related to input in terms of investment, hours worked daily and coverage of the ground; this is influenced by particular empirical needs for the task in hand, such as a sustained hard pull, long hours, high speed, etc. Figures for these are significantly affected by local conditions and by the condition of the animals; for example, at the start of the working season animals may take a few days to reach top performance, and individual animals may peak rapidly or maintain an unchanged work rate. Other factors can include the skill of the farmer, the training and condition of the animals, terrain, and rest periods. Nevertheless, I have made an attempt in Table 3.1 to provide comparative performance estimates and averages for ploughing donkeys and oxen from modern and other information. (The data are gathered from a wide range of specialist sources, too many to cite.)

Disease

In the dry climate of modern Iraq, endemic diseases affecting donkeys are not a significant issue aside from parasitic infestation and infected harness sores (Jaafar Jotheri pers. comm. 2014). In modern Lebanon, apart from some cases of glanders, donkeys mainly die of old age or accidents (Nasser Kalawoun pers. comm. 2017). Donkeys are well-known in modern Africa for being more resistant than cattle to and tolerant of a range of animal diseases, and for being able to cope better with parasite loads than many other domesticated animals; while parasites are commonly noted as a cause of death, this is likely to be through their low incidence of death from other

Table 3.1 Energy and work-rates for donkeys and cattle

Speed

- Ploughing speed (very much contingent on local factors): oxen 2.3–2.5 km/hour, donkeys 2 km/hour
- 'Equids are more suited to rapid low-draught activities where their faster speed can be used to advantage. At higher-draught forces, where speed is less important, the additional weight and power of cattle are an advantage' (Pearson and Vall 1998:309)
- Donkeys and female cows generally plough more slowly than oxen but are faster for lighter work such as seeding and weeding, and for transport; donkeys walk at c.4–5 km/hour
- Oxen work slowly, at a single speed, but have greater endurance in heavy traction; donkeys have a wider range of speed and better acceleration
- The extent of rest periods, for the animals and the ploughing individuals, is a factor
- A key factor in ploughing is the time that it takes to turn the animals at the headland
- Speed decreases if the number of animals in a team is increased: a trade-off is made between power and speed, and also working hours

Force/output

- Oxen are stronger for ploughing than donkeys in absolute terms due to their extra weight. A typical African ploughing donkey weighs 125–150 kg, and probably no less in antiquity (see Chapter 4.1 on donkey sizes), and an ox 300–500 kg, though oxen in ancient Mesopotamia are estimated to have weighed as little as 200 kg (Renger 1990:269)
- Absolute weight can be better for tasks such as threshing
- However, the extra strength is not necessarily required: farmers may shallow-plough, and the soil may be light and sandy, and not benefited by deep ploughing. The 2nd-millennium BC Sumerian text *Farmer's Instructions* (Black *et al.* 1998–2006d) recommends ploughing but is referring to a light ard, which would stir the earth to protect moisture but not dry out the ground as with a mouldboard plough
- Energy-rating studies conclude that donkeys can pull a larger percentage of liveweight than oxen (e.g. Dennis and Smith 1995:118; Prasad *et al.* 1991:236–7)
- Unsuitable harness such as a yoke has a significantly constricting effect on donkey output

Hours worked

- Oxen can typically plough for 4–5 hours/day, donkeys for 3–4 hours/day, though factors such as body condition can affect duration of work
- The Prasad *et al.* (1991:236) study cited previously asserts that donkeys can only work for two hours before becoming exhausted, but as discussed this relates to unsuitably harnessed animals. The authors themselves agree that it is very likely that use of a breastband would have increased the donkey hours worked, perhaps to equal those of cattle (Hagmann and Prasad 1995:237; Prasad *et al.* 1991:237)
- Unlike cattle, donkeys do not require a rest period during the day for rumination; they graze more slowly than cattle, but graze at night
- Oxen may set their own working limit, refusing to work harder, or may lie down and refuse to move, while donkeys can be induced to carry on working when exhausted (Lindblom 1931:22; Starkey 1989:164)

Ground coverage

- Accounts of ploughing with a pair of oxen and a simple plough indicate a normal coverage of c.0.2–0.4 hectares/day
- Donkeys will cover less hectarage not only due to lower speed and possible earlier tiring, but because they can turn more sharply and so plough more furrows per hectare
- Animal-powered weeding can cover more land than ploughing

causes. A modern issue which I don't address in this book is the major effect on use of working animals in humid regions of Africa, particularly cattle, of trypanosomiasis, carried by tsetse flies (see my Glossary in Appendix A). Taurine cattle (the species present in 4th- 3rd-millennium BC Mesopotamia) are mainly very susceptible, zebu cattle less so; donkeys are more resistant than cattle but do succumb – though the work-value of donkeys is such that farmers in these regions nevertheless use them, regarding them as a consumable.

Donkeys are also subject to colic if they are fed wheat grain or other rich food, and to harness sores and other work injuries due to their comparatively thin skin. In Iraq today, infection of harness sores is a common cause of donkey mortality, and in antiquity and today, unsuitable harnesses are used for donkeys in particular. Cattle have thicker hides, but Girsu texts from the 3rd millennium BC nevertheless attest to measures to prevent sores such as pieces of cloth put under the yoke (Renger 1990:273). Unlike horses, donkeys do not normally benefit from grooming: the dust content in their coats protects them from insects and heat.

Donkeys' hooves can become soft and overgrown if not used regularly on hard ground (e.g. Förster *et al.* 2013:204), or if donkeys are kept in damp regions. Both donkeys and oxen are susceptible to the hoof-damaging condition of laminitis; and in both, heavy work affects the hoof core, as noted in faunal remains such as in the donkey skeletons found at 3rd-millennium BC Tell Brak in northern Mesopotamia. Cattle are considerably more prone than donkeys to suffering incapacitating foot damage, particularly in drier regions and on hard or stony surfaces, and there are accounts going back to antiquity of concern for their feet. Old Babylonian laws and contracts mention penalties for damage to ox hooves (Roth 1980:129/132/135), and wrappings for damaged ox hooves are mentioned from Roman times to the present (e.g. Clutton-Brock 2003:126).

3.2 The poor man's ox: working cows

While the importance of male cattle for work in antiquity and today is undisputed, very little attention is given archaeologically to the role of working female cows. It is likely that the heavy association in Western countries of female cows with intensive dairying has contributed to this bias. The presence of castrated male cattle has been a traditional archaeological indicator of working animals (Piggott 1983:35), and this may well have contributed to the relative invisibility of working cows to date. Their common use in historical and more recent times underlines the need for a reassessment of models for use of working cattle in antiquity.

Sherratt (2006:336) refers briefly to the unsuitability of cows for heavy pulling in Neolithic Europe; but Gaastra *et al.* (2018:1473) offer evidence that light traction use of female as well as male cattle occurred from at least the 6th millennium BC in the Western Balkans. Cows have been used in the region throughout history, and most of the cattle still working in Europe today are cows (Starkey 2011:25).

Female cows are very widely used for work in Asia outside Hindu regions (e.g. Lawrence and Pearson 2002:103; Matthewman 1987:216–7; Starkey 1992:19), with a large proportion of draught animals in Indonesia and Bangladesh in particular being female. Minghao Lin (pers. comm. October 2015) notes that in north China in the 2nd millennium BC, cows were favoured for ploughing, and suggests that this was because bulls were often required for sacrifice. Working cows are common today in North Africa and Egypt; in an intriguing contrast, in Muslim areas in sub-Saharan Africa such as northern Nigeria and Ghana, there is notable reluctance to use cows for work. Elsewhere in sub-Saharan Africa, meanwhile, their use is common and increasing.

Use of cows for work is often associated with small farming units, particularly as if a single bovine is owned it will invariably be female. Kramarik (1975:132–3) demonstrates how the development in 18th-century AD Bohemia of harness suitable for single animals (unlike the traditional yoke) allowed small-scale farmers owning a single cow to utilise it for traction, helping to reduce the inequality that may have developed with early animal traction between those able and unable to invest in specialist draught animals, and reducing required herd size. There are advantages in putting male cattle to work to bulk them up for meat, but use of cows instead allows concentration on fattening the males and selling or slaughtering at the optimum stage, adding flexibility to herd and financial management (e.g. Matthewman 1987:219). In present-day Jordan, use of cows for work has been increasing as farming units become too small to justify investment in oxen, and in Africa the many official programmes to promote ox traction are sometimes sidestepped by small-scale farmers preferring the lower-key option of employing cows (Starkey 1994a:79) – or indeed donkeys. There are indications too from African experiences that women find cows more acceptable than oxen to use for work, making cows an attractive choice for a small farming household.

Cows may be easier to train but are generally smaller and need to rest more often; they are not necessarily less strong, weight for weight, than oxen, and the extra absolute power of oxen is not always needed. Views vary on whether cows (and uncastrated bulls) live and work longer than oxen; Halstead (2014:16) reports that in recent Greece working cows became worn out at an earlier age than oxen, while others note that working cows live to perhaps 15 years or more. The difference may lie in the greater focus on the meat value of oxen: Reh (1982:80) reports that in Africa

> cows can be used for draught work over a lengthy period, thereby reducing the need to train animals so frequently (7–9 years in comparison with 2–5 years, particularly if the draught oxen are still to achieve a good meat value).

Commentators' views vary on the extent to which moderate working can affect a cow's ability to produce calves and milk; with sufficient extra feed – more than for an ox – the net benefit of combined work, calves and milk may well be greater than that of using an ox. The regime for working cows, including their unavailability for work before and after calving, does require additional planning, but unlike donkeys' seasonal oestrus, cows come regularly into oestrus year-round. The close monitoring and stall-feeding of animals on an intensive smallholding would indeed provide the high level of feeding and care needed to allow a working cow to maintain milk and reproduction levels (e.g. Halstead and Isaakidou 2011:62/66; Prasad *et al.* 1991:234).

Archaeologically, evidence of the use of working cows might therefore indicate the presence of advanced agriculture on small independent farms; this may have occurred early on in Egypt, where illustrations indicate use of cows from the 3rd millennium BC. In Neolithic Crete, mainland Greece and Germany, evidence of traction pathologies in cows, but not male cattle, may indicate small-scale cultivation, with Isaakidou (2008:109) even suggesting a model where the adoption of cows for work preceded that of male cattle. The evidence over time of adoption of cows for work, if modern parallels with Africa can be used, may also be an indicator of agricultural change of various kinds. If farmers are short of working animals due to disease, drought or increased demand for crops, use of working cows may increase. Alternatively, farmers may put their cows to work if there is a shortage of investment capital, grazing land or fodder and they wish to minimise their overall animal holding: use of an ox tends to be seasonal whereas a cow earns its keep year-round.

So reduction in herd size may – confusingly for archaeologists – be a cause or an effect of using working cows. There are of course arguments for large herds unrelated to work needs if the investment capital is available; reducing herd size by removing males and relying on working cows does expose the farmer to simultaneous loss on several fronts if a cow is ill, dies or is stolen.

As with donkeys, this concentration of working cow use among small-scale farmers may have largely shielded it from historical recording; Johannsen (2011:14) suggests that

> [t]he use of the stronger oxen may have been the practice that was regarded most highly and therefore the one given most attention in writing and iconography.

The growing zooarchaeological evidence of working cows – whether for sledges or ploughing – underlines the need for a reassessment of models for use of working cattle in antiquity, as voiced by Johannsen (2005:47):

> it is necessary to consider the variability of modern and historically known draught cattle husbandry strategies when studying the prehistoric use of draught cattle, and especially when evaluating mortality patterns.

Female cows may largely have been confined to the breeding (and presumably dairy) centres: Maekawa (1979:102) reports from pre-Sargonic (mid-3rd millennium BC) Lagash texts that only males were supplied from these centres for ploughing. The 3rd-millennium BC Ebla texts (Pettinato 1991:94) refer only to male cattle for work; in some centrally controlled agricultural regimes the extra feeding and care needed if cows are to work, as well as provide milk and calves, might not be considered worthwhile. By the late 3rd millennium BC, though, lists of ploughing animals from Girsu and Umma record that almost a quarter of centrally-controlled ploughing cattle were female (Heimpel 1995:89), and there are early 2nd-millennium BC Old Babylonian accounts of cows working at ploughing and threshing (Stol 1995:176/185/188–91). It is possible, though, that non-working females producing calves for a plough team are included in the listings:

> The purpose of the cows was to keep the number of the draught animals up to strength with their offspring. . . . However, in the absence of adequate numbers of oxen, cows could be used for pulling the ploughs. Cows trained for this task were designated as šaddītus.
>
> (Jankovic 2013:51)

A note on the use of female donkeys

Pre-Sargonic Lagash (mid 3rd-millennium BC) texts from Sumer record that female donkeys were mainly employed for ploughing and males more for drawing vehicles (Maekawa 1979:102/111); this may reflect the greater strength or prestige of males, or the desire to keep males and females apart to avoid bickering among the males in the presence of females or to prevent breeding: Maekawa suggests that some form of spaying or breeding control may have been exercised on some females. In modern India, carting is similarly assigned to males, and pack to females. A final hypothesis might relate to the seasonal nature of ploughing, with females then available for the rest of the year for reproduction. This regime seems to have altered by the late 3rd millennium BC (or geographically), when texts from Girsu and Umma indicate that at least half of donkey ploughing teams were male (Heimpel 1995:89/96). In texts on the early 2nd-millennium BC Kaneš pack caravan trade there is only rare mention of females, perhaps because females are occupied in reproduction, as at 3rd-millennium BC Mari where texts from a breeding/supply centre refer mainly to females (Sallaberger 2014:341/347). In modern times where there is a preference for male donkeys for work, females are concentrated in remote breeding areas, leading to a circular issue in which donkey users must obtain working donkeys from the specialist breeders as they do not have easy local access to breeding females.

While male donkeys may display aggressiveness or attempt to mate, experience in modern donkey-using regions worldwide demonstrates overwhelmingly that mixed groups of male/female donkeys generally operate without trouble. While there is sometimes prejudice about the comparative work output of males and females, other users confirm that performance is similar, allowing for size. Some reduction in work output occurs in pregnant females, but they can work until an advanced stage. In many donkey-using regions females are widely employed, with little differentiation in choice unless the owner is planning to breed; however, there are significant cultural variations, sometimes even within narrow geographical areas.

4 Working-animal supply logistics

4.1 Breeding and supply of working animals

Introduction

The 'shift of focus from the dead to the living animal' (Meadow 1984:310) with the advent of working donkeys and cattle brings some very different practical and economic considerations into human life. I concentrate particularly here on the breeding and supply of donkeys as a species mainly raised for work. Despite more donkeys today than ever in prehistory being bred and traded, there is little published material to assist us in reconstructing likely systems in ancient Mesopotamia and beyond. Breeders and traders today, as in antiquity, inhabit remote regions and prefer to keep their activities out of official records, and as profitable entrepreneurs they are not the targets of aid organisations. Dercksen (2004:258) underlines the scarcity of provenance information in 2nd-millennium BC texts, very possibly for similar reasons, on the huge numbers of donkeys used in the caravans described in the Kaneš texts.

The modern West has lost touch with the complex nature and central importance, until recent times, of working equid breeding and trading, equivalent perhaps to the vehicle industry today. McShane and Tarr (2006:370) describe the system 150 years ago in the USA for supply of horses to urban areas:

> [a]n elaborate market assigned value to animals. Breeders sold horses to owners of feeding lots. The feeders in turn shipped horses that they had broken to harness to central markets in Chicago and Kansas City. From there the horses went to wholesale auction houses in large cities. . . . Secondary markets specialised in used horses, and there was much informal trading comparable to the current market for used cars.

The evidence from both antiquity and modern analogy is that large-scale breeding of donkeys for work, from the time of their early systematic use in Mesopotamia, has been conducted by specialists and located in remote, arid regions suited to their natural habitat. The transfer of animals to areas of demand, from early times to the present day, has arguably always been conducted through chains of regional livestock and goods markets which form a crucial interaction point between mobile groups, who in turn have productive contact with sedentary cultures.

Unique features of donkey-breeding

The breeding of donkeys differs essentially from that of meat animals in its aims and process, not least because domesticated meat species are generally herd animals: donkeys are sociable in the wild, but congregate in small, shifting groups rather than herds. Reproduction in donkeys is slow: female donkeys may not foal until the age of four or five in hot climates, and typically only produce a foal every other year, partly because oestrus tends to be seasonal and because females do not 'show' during oestrus so may miss mating, especially in the low-management systems typical for donkeys. Mixed groups of working donkeys from different locations often graze communally, leading to unselective breeding. Up to modern times mobile groups in northern Africa have also caught and tamed wild donkeys for transport, and actively encouraged impregnation of females by local wild stallions, to provide renewed genetic vigour and the continued ability to endure disease and poor feed. Shackelford *et al.* (2013:4171) add that

> [t]he deep roots of this tradition are further illustrated by the first century CE agronomist Columella, who discussed the benefits of backcrossing female donkeys with male wild asses in his treatise De re rustica (6, 37, 3e 7; English translation 1745:298e301). Together these factors point to relatively low levels of selection and contextualize the lack of consistent size decreases in early domestic donkeys.

The recent unlocking of the mitochondrial DNA of modern domesticated and wild donkeys gives clear evidence of long-term gene flow between the two; this evidence has advanced the debate on original domestication location and timing, indicating more than one independent domestication episode, but can also support the hypothesis – abundantly demonstrated in modern and historical times – of deliberately permeable boundaries between wild and domesticated donkey populations on a day-to-day basis (Larson and Fuller 2014:118).

The zooarchaeological, modern analogical and genetic evidence indicate strongly that aside from specialist centres donkeys have never been widely subject to strategic breeding – though the culling/castrating/selling of males (to remove poor specimens or achieve profit) will have had some effect. Any reproductive isolation of non-specialist donkeys in antiquity is likely to have occurred in urban settings of donkey use, for example in late Predynastic and Dynastic Egyptian towns (Marshall and Weissbrod 2011:S407), or in the very controlled environment of 3rd-millennium BC Mesopotamian agriculture, at which period there is evidence on equid figurines of features such as genital strapping to prevent breeding, and of oestrus, indicating interest in guiding the breeding process (Zarins 2014:93/136–7; Hauser [2007] 2008:469 in Figure 4.1).

Breeding zones and long-distance transfer

There are accounts throughout history of some regions known for their suitability for breeding particular animals. Animals have natural ecological niches, especially in arid areas; donkeys (*Equus asinus*) are native to the rugged, dry, often hilly terrain of North-East Africa, and in modern times – as very probably in antiquity – their large-scale breeding zones reflect this environment. There can be good reasons to move animals from their natural habitat: prices for the animals will be higher elsewhere, and there will not be wild herds to tempt the domesticated animals away – a common issue to the present day with reindeer and camels.

In modern West Africa, donkeys are bred in large numbers in arid Saharan and northern Sahelian countries such as Niger and northern Mali by specialist groups such as the Teda, and are supplied profitably to more humid regions further south, where donkey fertility is lower and where humidity-related diseases, unsuitable foodstuff and cultures unused to donkey management lead to short lifespans. Donkey merchants and traders from the south travel frequently to the huge markets in the north such as Djibo, Gorom

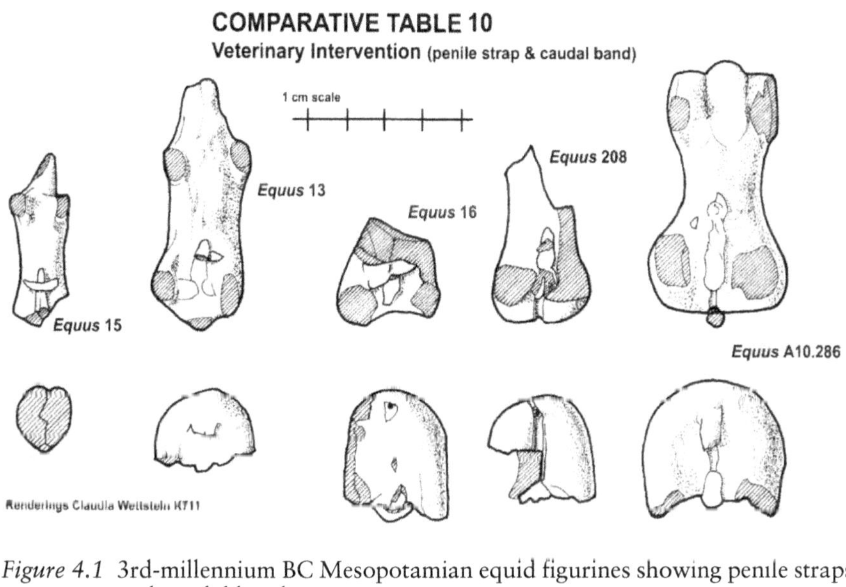

Figure 4.1 3rd-millennium BC Mesopotamian equid figurines showing penile straps and caudal bands

Source: Hauser [2007] 2008:469, Table 10, by permission

Gorom and Dori to buy consignments of donkeys for central Burkina Faso (JG Burkina Faso 2013), and 'Hausa and Zarma traders buy the donkeys directly from the producers and take them to the border markets where they enter the Nigerian trade system' (Blench *et al.* 2004:217). A similar chain-of-supply system exists within Ethiopia (e.g. Pearson *et al.* 2001:64/67); my interviews with dealers and buyers in north-western Ethiopia (JG Ethiopia 2014) found that individuals in drier regions such as South Gonder breed a few extra donkeys for income and sell unneeded animals at local weekly markets. These are bought by traders, who collect a group of donkeys and take them south-westwards on foot up to 200 km to humid, tropical regions of high demand. The donkeys are purchased at the frequent markets in the demand-area villages (Figure 4.2) by individuals replenishing plough or pack animals. The journeys with the donkeys are dangerous ones due to predators and thieves, and there may be feeding issues (Pearson *et al.* 2001:70); young donkeys are then sold for pack, older ones for ploughing. Some farmers sell their ploughing donkeys at the end of the season and re-buy next year, but in many cases the donkeys die of trypanosomiasis and other humid-area diseases within a couple of months of purchase; many of the owners seem to accept this high loss rate (Starkey 2011:25) and regard the donkey as a consumable, as unfortunately with slaves historically in America and elsewhere.

Förster *et al.* (2013:197) also describe large consignments of donkeys in the present day taken 900 km by donkey buyers from Debba in northern Sudan – where the donkeys are 'famous for their strength and endurance in heavy work and carrying loads over long distances' – to the El-Fasher

Figure 4.2 Donkeys from Gonder being sold at the thrice-weekly market in the village of Jawi in humid W Ethiopia

Source: © JG Ethiopia 2014

market in Darfur to the west, where the drivers can sell them at three times the purchase price. Consignments of donkeys are also collected and driven similar distances for *ad hoc* demand, as when a tribal group near Timbuktu had lost most of its donkeys and camels to drought and civil war; a tribal group at Bouressa (an area with an abundance of free-living donkeys) drove a consignment of donkeys to Timbuktu for a fee (Förster *et al.* 2013:209). Conflicts can have a significant effect on donkey populations: in modern Mozambique after a series of debilitating wars, farmers lobbied the government for imports of donkeys from neighbouring Swaziland (Clarence-Smith 2016:120).

In Namibia in the mid-20th century AD use of donkeys was officially discouraged by authorities, who favoured oxen, so donkeys were purchased in the markets in the more temperate south and driven 'in herds on sometimes months-long drives to Ovamboland' in the north (Kreike 2010:98). With growing demand for donkeys in post peak-oil Africa, donkeys are also being increasingly supplied across national borders by countries with established donkey populations to neighbouring regions who lack a tradition of using donkeys. This process may overlap with theft: Hagmann and Prasad (1995:236) report, for example, on high levels of donkey theft in Zimbabwe, a surprising incidence given that oxen are widely preferred there for work, and donkeys are subject to prejudice as reported by a number of regional experts; but a likely explanation is that donkeys are being stolen for supply to neighbouring countries. Large-scale animal-rustling has a long history: Native Americans, for example, found horse-raiding easier and more prestigious as an occupation than horse-breeding, despite the good conditions for the latter.

The history and detection of markets

In 16th/17th-century AD England, horse-dealing fairs were commonly situated 'at the junction of breeding and rearing areas' (Edwards 1983:121), though generally the buyers had to travel further than the breeders (p. 126). In developing regions using large numbers of donkeys today, donkeys are commonly traded at huge regional markets near to the breeding zones, attended by traders and donkey merchants from demand regions who convey the animals back and sell them at smaller markets or retail; payment may be in goods such as grain. These large markets, selling a range of working and meat animals, have their roots in antiquity, both in Africa and in Syria, Pakistan, Baluchistan, Afghanistan, India, China (Figure 4.3) and elsewhere. Their presence, as fixed periodical meeting-points for goods transfer, may also be assumed in early times, for example on the Iran plateau trade routes for obsidian, copper and semi-precious stones. In areas of high demand, traders may bring donkeys to regular markets, as described previously in Ethiopia, or to an urban 'pop-up' market, as is common in many parts of modern Africa in particular, where on a known day, the streets of a town

Figure 4.3 Donkey section of Kashgar market, Xinjiang, China
Source: © Jill Goulder 1991

fill with donkeys brought in for sale (JG Burkina Faso 2103; Paul Starkey pers. comm. 2015). There are indications that a similar system operated in Mesopotamia in the early 2nd millennium BC, where an Old Babylonian commercial letter advises a recipient 'concerning asses that you need, come here and buy asses, the asses have come up from the country' (Tablet BM 97347, CT 33 21).

The traces of ancient and even modern markets, sometimes in remote areas, are difficult to detect except by traded material distribution patterns, though modern archaeological investigation can be aided by satellite imagery and isotopic analysis of dung presence. Archaeologists may underestimate the importance, in terms of exchange processes and of livestock producers' strategies, of these gatherings; to this day they are central to livestock-rearing economies, aiding efficient circulation of unneeded animals such as excess males, and they are as much channels for diffusion of ideas and skills as of animals.

Specialist breeders

In contrast to the plentiful early textual mentions of professional donkey herders (Zarins 2014:230–1) there are few references to breeding apart from semi-nomadic Haneans breeding donkeys locally, as in early 2nd-millennium BC Mari texts (Kupper 1957:6–15; Zarins 2014:201/249). While the accumulation of donkeys at regional centres for onward transfer was susceptible to state control, I suggest that the systematic breeding sector itself – perhaps concentrated in remote regions – was likely to have been privately operated. As I discussed in Chapter 2.2, Zarins (2014:109/201/249) and others suggest very plausibly that mobile pastoralists were responsible for the introduction of the domesticated donkey to sedentary and urbanising groups in the Ancient Near East; but as donkeys are not commonly eaten, they are not an asset to be taken in large herds on pastoralists' long seasonal rounds. I suggest that when systematic use of donkeys evolved in the late 4th/early 3rd millennia BC, the breeding of donkeys then became a specialist business for certain less-mobile but still non-urban sub-groups. Milevski (2011:177/196/232–5, discussing 3rd-millennium BC southern Levant) also suggests from modern analogy that donkeys at the time were likely to have been bred by specialists, as in modern Arabia where the Solubba breed donkeys for other groups.

There are numerous historical examples of specific groups specialising in breeding or dealing in working equids, with 'gypsies' famously specialising in horse-dealing in Britain. Symbiosis is sometimes an element: in 16th-17th-century AD England, itinerants related to small-scale horse-rearing farmers dealt in horses for urban areas (Edwards 1983:114–16), and in 18th-century AD North America, village inhabitants were kept supplied with horses by related nomadic groups (Ewers 1955:19–14).

Equid supply flows in Mesopotamia

References to donkey supply in the 3rd- and 2nd-millennium BC Near Eastern texts underline the large-scale geographical complexity of the equid-breeding industry then as now. The 3rd-millennium BC texts from Lagash and Mari indicate donkey (and later horse and hybrid) supply centres at Dêr and Gutium in the Zagros region, Mari, and Kish in central Mesopotamia (Sallaberger 2014:350; Zarins 2014:160/199); a tablet from Tello (ancient Girsu) refers for example to more than 700 donkeys transferred hundreds of kilometres from Gutium to Lagash (Figure 4.4). The Zagros centres specialise in working equids, in contrast to the major meat-animal distribution centre at demand-area Drehem in the later 3rd millennium BC, where there is little mention of equids.

Kupper (1957:36) then describes early 2nd-millennium BC accounts of horses being bred in southern Syria, traded in Qatna and provided to areas further north, as in the Letter from Bahdi-Lim to the Hanean king Zimri-Lim; this strongly indicates development of new desirable breeding regions distinct from indigenous horse regions to the north-east. Mapping the locations of working-animal transfer activities, then and more recently, can provide clues to working-animal supply and demand systems and associated large-scale activities perhaps otherwise difficult to detect.

References to quality of donkey stock, and to knowledge of the best regions, also appear in 3rd-millennium BC Ancient Near Eastern texts from Ebla and elsewhere on the hybridising of donkeys with onagers: good donkeys were imported to breeding centres in or close to regions with indigenous wild onagers, in northern Iran and further west as with Hamazi in Iraq and Nagar (Tell Brak) in north-eastern Syria (Archi 1998:10; Zarins 2014:171).

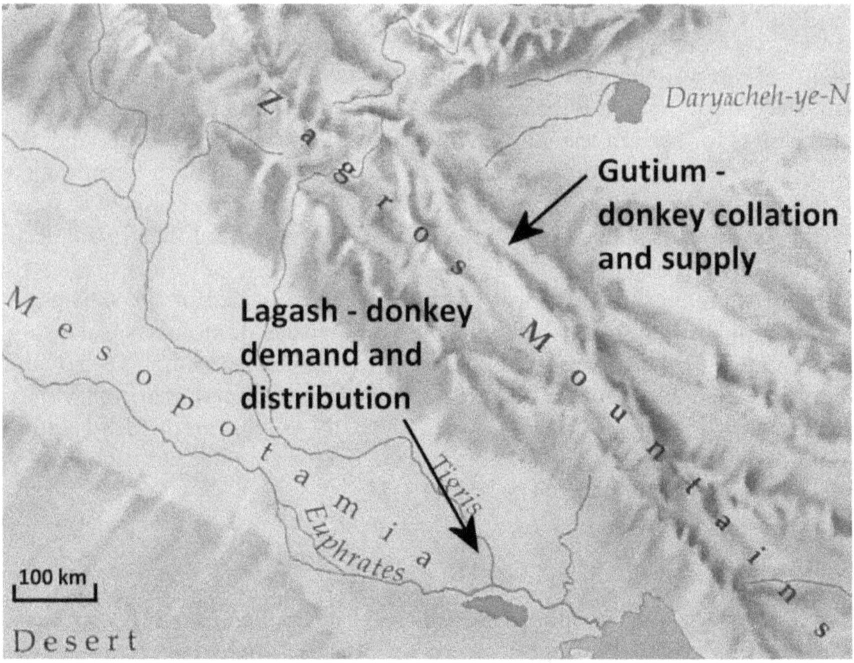

Figure 4.4 Late 3rd-millennium BC donkey supply and demand

Source: Based on extract from Times Atlas of the World 1995:43. © Collins Bartholomew 1995, by permission

Heimpel (1995:91) reports on late 3rd-millennium BC texts recording specialist suppliers of wild male onagers, evidently for breeding. In early 20th-century AD Iraq, mules were still bred in Iran and Kurdistan and prized for their superior quality (Admiralty War Staff 1916:139; Zarins 2014:251–3).

Zarins (2014:245) summarises the textual references to donkey supply and demand in 3rd-millennium BC Mesopotamia in terms that closely describe the modern African situations described previously:

> mortality rates were high and longevity was at a minimum. Therefore, the net local effect was negative, and, as a result, constant purchase from the peripheral areas of Mesopotamia was necessary to supplement local breeding.

Brewer *et al.* (1994:99) argue similarly that 'donkey breeding in Mesopotamia was still relatively unsuccessful and stocks were replenished through trading (and raiding) with northern and eastern areas', though this erroneously implies incompetence rather than a well-organised system of breeding of disposable donkeys in the most suitable regions, as today, with users finding economic sense in buying healthy donkeys rather than attempting to breed them where a suitable environment or space away from farmland is available.

Donkey categories

Although, as I describe earlier, the majority of donkeys have always been the result of unselective reproduction, there are, meanwhile, many modern and historical examples of specialist strains of donkey developed for particular purposes; Darwin in 1868 AD noted four breeds of donkey, used for different types of work. Donkeys from certain regions are valued as hardier, larger or stronger, and in many cultures there are different breeding aims for riding and for pack donkeys. A specialist in breeding or supply of 'big donkeys' is mentioned in 3rd-millennium BC texts from Mari (Sallaberger 2014:341); and in early 20th-century AD Iraq two categories of donkey were bred: for riding (large, strong, handsome, expensive) and for pack ('black, and dark brown, and is small and wiry in appearance' (Admiralty War Staff 1916:139)). In modern Egypt, Sudan, Ethiopia and further south, the preferred riding donkey is larger, stronger and faster than the ordinary pack donkey, though less enduring and hardy.

References to donkeys in the early 2nd-millennium BC Kaneš texts concerning the donkey-caravan trade from Aššur commonly include the adjective 'black', particularly when attached to a detailed list of goods being conveyed, such as '1 talent and 6 minas of tin and 5 black donkeys' (Barjamovic 2011:88). What were the attributes of these black donkeys? There are occasional Kaneš text references to demand for large, strong and young donkeys; certainly strength is required, but very large size may not be an asset in pack donkeys, which need to be short-legged enough for easy loading and presumably catching, though long legs are useful for difficult terrain and for example fording rivers. Dercksen (2004:258) notes that 'the colour is the sole characteristic of these donkeys which OA texts contain', and it may be that the 'black donkeys' specified are not special ones but the opposite: the writer is indicating that the donkeys listed are common pack donkeys rather than expensive riding animals.

There is a general tendency for size reduction in domesticated animals through lower access to nutrition and exercise than in the wild or for reasons of stress or breeding in overstocked conditions. There may be active selection for small, more manageable animals for meat, milk or wool/hair, and smaller animals may be more likely to survive than in the wild among predators. However, among working animals there may be active selection for large size, and changes in size of faunal remains compared with the wild may also reflect hunting strategy regarding the former, or environmental factors or episodes of genetic isolation. The size-reduction criterion is nevertheless employed or referred to by quite a few zooarchaeologists and other commentators investigating early donkey finds in the Ancient Near East; but any size reduction over time in domesticated donkeys in the Ancient Near East appears in fact to be slow and non-linear (Marshall *et al.* 2014:6154; Shackelford *et al.* 2013:4170–1). This lack of overall size reduction may of course relate to the elite status of the donkey specimens examined (though see Chapter 2.2 on 'elite' donkey burials), or more practically to the low-management regime of donkey-keeping and continued recruitment from the wild that militates against domestication features seen in other species.

Grigson (1993:645–6) cites reported evidence of domesticated donkeys eventually reducing in size, but (2003:219) notes the uncertainty of this criterion, recognising the desirability of such an indicator as (2003:220) '[t]here are no other really reliable anatomical criteria for the distinction of wild and domestic equids'. Shackelford *et al.* (2013:4170–1) and Vila (2014:434) report that some early examples found are indeed smaller than their wild equivalents, but the donkeys with work pathologies buried at early 3rd-millennium BC Abydos in Egypt are as large as their wild equivalents, as are examples from the later 3rd and 2nd millennia BC in north-eastern Africa and at Tell Brak in Syria. Dercksen (2004:257–8) records the heights of donkey remains (some in ceremonial contexts) from 3rd-millennium BC Mesopotamian sites as ranging from 109–129 cm withers height, with a slight decrease over time to the donkeys at late 3rd-millennium BC Tell Brak (109–115 cm). Clutton-Brock and Davies (1993:210) report on the Tell Brak finds that these donkeys therefore differ little in height from 'the short-legged donkeys of the present day [in Syria] that commonly stand about 11 hands (112 cm) at the shoulders', setting aside taller specialist riding-donkeys. Modern African donkeys are not habitually measured, but stand typically at c.105 cm (e.g. Wilson 1978:185).

Breeding of donkeys for caravans

Animal owners supplying the caravan industry (with donkeys, reindeer, camels, yaks or llamas) may either sell or hire out their animals to merchants, sometimes with themselves as drivers, or they may organise caravans themselves for the traders. These systems could be categorised respectively as symbiotic and integrated, and their incidence may reflect the original impetus for caravans – 'push' by breeders adding value by transporting goods, or 'pull' by entrepreneurs requiring transportation services. Joffe (1991:28) suggests that exchange in the 4th millennium BC between the southern Levant and Mesopotamia was 'mediated over great distances across arid zones by nomadic or semi-nomadic groups', with or without donkeys – the integrated system – though Oren (1989:401) suggests that in northern Sinai this process may have been conducted by the town-dwellers, implying that the nomads were carriers and donkey drivers working symbiotically with merchants.

I cite Milevski (2011:177/196/232–5) previously as suggesting separate donkey breeders and caravan-operators in 3rd-millennium BC southern Levant; he goes further to propose that either or both categories may have formed socially-separated groups with donkey-related cultic adherences manifested in donkey figurines and/or elite donkey burials.

The system used could be a valuable pointer to the overarching political structure of the period, as it could be argued that in authoritarian regimes, responsibility for caravans is more likely to lie with animal-owners than with independent merchants. Earlier interpretations of 3rd-millennium BC Mesopotamian exchange activity assumed state control, but there is a growing body of evidence from texts and archaeology that there was also privately organised long-distance trade by self-sufficient communities. There is almost no reference in 3rd-millennium BC official texts to what must have been a significant process of supply of donkeys for long-distance caravan work, apart from a passing mention in a Jemdet Nasr-period text (IM 23435) of 'travel donkeys' as distinct from donkeys for ploughing (Zarins 2014:199). This contrasts with the common official listings that I note earlier of transactions concerning donkeys for agricultural work.

I outline earlier how donkey breeders in modern Africa raise their animals in remote, arid regions and sell them in a continual flow via large markets to donkey merchants, who keep smaller groups of donkeys near urban and agricultural areas, supplying individual farmers and transporters and occasionally replenishing caravans. Meanwhile modern merchants organising caravans may buy in donkeys from breeders, as for the early 2nd-millennium BC Aššur-Kaneš caravans, or sometimes breed their own; the dual outlets for donkeys – to donkey merchants for local employment and to caravan-organisers – help to make large-scale breeding of a non-meat animal logistically viable. In the Kaneš system, donkeys were also hired in at times, though this may be from local owners rather than the breeders themselves; and there are accounts of donkeys being given and received as payment, often as part of the 'package' including the goods. There is little textual information on the provenance of the Kaneš donkeys, with only isolated references to locations for donkeys being bought and sold, for example at a market at Šinahuttum in eastern Anatolia and at gigamlum, paddocks and breeding centres near Aššur (e.g. Barjamovic 2011:43/288).

As well as pack animals for the start of a caravan journey, a supply of replacements was and is still required along the route to deal with erosion of animals due to illness, mortality, uncooperativeness, escape and theft. The Kaneš texts report high donkey loss rates from the caravans from Aššur, replenished by buying or renting *en route*, and fresh donkeys had to be found for each new caravan as most donkeys were sold in Anatolia on arrival (Dercksen 1996:61). The donkey sale price in Anatolia equalled the purchase price, perhaps due to the ex-caravan animals' proven strength and to demand, though sometimes there must have been an excess of donkeys on the Anatolian market. While a few equid bones have been found in the

Anatolian area of Kültepe, possibly indicating some use for meat, the great majority must have been sold for local work.

Breeding of cattle for work

In contrast to donkeys, cattle have throughout history been widely kept for meat and milk, and so the breeding, buying and selling process often relates to this rather than to work use. Ur III and Old Babylonian texts from the late 3rd/early 2nd millennium BC describe cattle-supply stations at Sippar, Kish, Larsa, Girsu and Ur (Heimpel 1995:110; Stol 1995:180), with indications of a focus on breeding for work strength (Stol 1995:184). A revealing slant on the recording practices of the late 3rd millennium BC is given by Liverani and Heimpel (1995:128–9), re-examining an Ur III tablet listing annual figures for growth of a herd of cattle over ten years and finding them wholly unrealistic in annual terms: as the administrators do not have access to the actual figures, they are providing a statistical model which provides a reasonable breeding 'goal' figure at the end of ten years.

A potential issue with supply of working cattle is the division found at times in Africa between cattle-herding and agricultural work, but at a village level, farmers might gain income from buying, bringing on and selling oxen. There can be an element of working-animal breed specialisation: in south-western Burkina Faso, farmers own small taurine cattle for meat, milk and prestige but buy in zebu (locally considered stronger for the heavy soil and hilly terrain, albeit less suited to the chilly, damp climate) for ploughing (JG Burkina Faso 2013).

There is, though, a long history of specialist breeding centres. Cattle, as with donkeys, flourish in certain environmental conditions and not in others, for reasons including disease and lower fertility, so in many regions of modern Africa and elsewhere, there have traditionally been breeding centres from which cattle are supplied to other regions via traders. These might prove difficult to detect archaeologically, though Redding (1992) gives a useful account of the zooarchaeological identification of a 3rd-millennium BC Egyptian cattle-breeding centre by the low incidence of cattle bones (notably those of prime age) combined with the high presence of cattle dung.

One system offering potential parallels for investigation in antiquity is a cattle-supply system in the 19th century AD in Ethiopia, in which farmers in the north-eastern highlands benefited from an elegant multi-point arrangement for maintaining their ox-based plough agriculture, involving donkey caravans (McCann 1984:5–10; Pankhurst 1968:241/346–56). The Ethiopian highlands are fertile but feature heavy clay associated with steep slopes, workable by oxen but not easily by equids; however, the climate limits the reproductive capacity and lifespan of cattle. Oxen were vitally needed as the continuation of annual cultivation was essential, under tenure arrangements, for farmers to maintain their land. The Lasta highland farmers were able to buy cattle for work at the weekly market at Saqota, purchased by Saqota merchants taking Danakil salt south-west to markets such as Ibnat. Merchants from the cattle-breeding areas in Gojam to the south-west brought cattle, cloth and coffee to Ibnat to exchange for salt, and the cattle were then taken to the Saqota market to sell to the farmers. The salt was brought up (as to this day) by donkey/mule/camel caravan from the Danakil depression in the east to Mekele in Tigray, and on south-west to the Saqota market. The system (which uses no facilities that were not available in 3rd-millennium BC Mesopotamia) was huge in scale – there are reports of salt caravans of 15,000 donkeys, and of 21,000 cattle/week exchanged at Saqota – and stretched hundreds of kilometres.

4.2 Donkey meat, hides, milk and dung

Eating donkeys: cultural taboos

One of the fundamental drivers for writing this book is the issue of under-representation of donkeys in the faunal record, in consequence of the common (albeit not universal) cultural avoidance of consumption of donkey meat from antiquity to the present day. Food-middens are typically our main source of archaeological information on animals in ancient cultures, including on working cattle, which are normally eaten after their working life is over. There has been extremely little recognition to date of the invalidity of applying the same approach for donkeys.

Modern analogy supports the hypothesis that the remains of non-food animals are unlikely to be found in settlements: dead donkeys may be dragged well out of camps and villages, and donkeys no longer able to work are abandoned in the bush/desert rather than killed (e.g. JG Burkina Faso 2013; Marshall and Weissbrod 2009:72); donkey remains are rare in settlement refuse dumps. Donkeys' transport work also takes them far from settlements, and when they die in harness, their abandoned carcasses are likely to be consumed by predators. Brodie (2008:304) comments that

> [t]he absence of donkey bones from domestic assemblages of what were probably food remains might say more about dietary preferences than about the use of the donkey as a work animal.

Where the first osteological traces of domesticated donkeys are identified, at 4th-millennium BC and later sites in the Ancient Near East, the indications are that they were not generally eaten, as aside from the phenomenon of elite burials of equids, they appear to have been dumped whole, as at 3rd-millennium BC Abu Salabikh in Mesopotamia, and without cutmarks except occasionally for skinning for hides (e.g. Payne 1988:99–103; Vila 2006:103/108). In contrast, the bones of food animals are commonly found broken and scattered, with a concentration of meat-bearing bones, and show cutmarks and evidence of cooking. Payne (1988:99–104) reports on the contrasting state of onager and donkey bones at 4th-millennium BC Tell Rubeidheh in Mesopotamia; the remains of the indigenous onagers, commonly hunted for meat, are fragmented and scattered, unlike those of the donkeys, which were not indigenous and constituted valuable assets.

A potential early exception has been recorded at 5th/4th-millennium BC Tell Arjoune in Syria, where equid bones (probably donkey) were found in a broken and scattered pattern (Grigson 2003:220), while at 3rd-millennium BC Khirbet al-Batrawy in the Jordan valley there are possible signs of donkeys being consumed (Alhaique 2008) – perhaps pack donkeys at the end of their lives – though, as always, the picture is blurred by the presence of wild onagers. In a later period, Michel (2004:203) reports possible zooarchaeological evidence of donkeys consumed at early 2nd-millennium BC Kaneš, where large numbers of superfluous donkeys would have accumulated from the Aššur-Kaneš caravans, obviating the practical need for taboos.

Vila (2006:103) and Zarins (2014:202) report no epigraphic evidence of donkeys being eaten in 3rd-millennium BC Mesopotamia, though donkeys are reported as fed to dogs and possibly killed to seal the forming of an alliance, as mentioned in a text from Mari. In the records of Drehem, a major Mesopotamian animal redistribution management centre in the later 3rd millennium BC, donkeys are mentioned but in comparatively small numbers. This may reflect their smaller Mesopotamian population compared with the large herds of cattle, sheep and goats, but it may be significant that unlike these food animals, the carcasses of dead donkeys are sometimes received, and that donkeys are recorded as being destined for feeding to dogs and captive lions (e.g. Zarins 2014:202/251); this may indicate that donkeys passing through Drehem are not intended for human food.

It is now generally concluded that the 3rd-millennium BC burials of donkeys in elite human graves are not related to food offerings, though Way (2010:211–15) leaves the question open on later examples in the southern Levant. In ancient Egypt there is occasional mention of donkey meat, and parts of donkeys are recorded as being used medicinally, but the meat then appears to become taboo, perhaps by reason of the donkey's association with the god Set (e.g. Osborn and Osbornová 1998:136); Marshall (2007:384), though, suggests that the burnt donkey remains at 3rd-millennium BC Maadi might relate to use as food.

In modern Africa there are indications that cultural taboos on eating working equids still apply to domesticated donkeys in many situations outside emergencies. In a large body of overviews of donkey use by regional experts and findings from specific studies initiated by government organisations, published in academic journals and official aid organisation documents (too many to list here, but see, for example, Blench 2004:22–4; Blench *et al.* 2004:213), the eating of donkey meat even in regions of major use is confirmed as comparatively rare and largely confined to specific minority groups. Traditionally in these groups there is low-scale eating of culled and aged donkeys, with only occasional instances of deliberate fattening of donkeys for consumption, though the eating of donkeys in Africa is probably more widespread than is commonly reported, and is increasing in some areas. A movement of (older) donkeys between non-eating and eating groups, as from Muslim to Christian and animist areas (William Clarence-Smith pers. comm. 2016), may be zooarchaeologically detectable.

Donkey meat is tough, and the flavour is disliked by some, so is often found smoked or processed into salami. Outside Africa, donkey meat is reported as eaten in Greco-Roman times (Bodson 1985:6; Mitchell 2018:122) and in modern Europe (e.g. Audiot and Garnier 1995:66), as well as increasingly in China. Taboos may be born of practicalities – the donkey is long-lived, and there is no economic sense in killing a trained, experienced animal while it is still useful. Lattimore (1928:144) notes the different behavioural codes relating to camels in Mongolia, between cameleers who had strong taboos about killing their working camels (though will eat wild ones), and traders who freely ate camels. Similarly, Native Americans with small numbers of horses were generally reluctant to eat their working animals while those owning large horse-herds and with a less close working relationship would eat horses (e.g. Ewers 1955:222). In Africa, Marshall and Weissbrod (2009:71) report Kenyan Maasai pastoralists' objection to eating animals who work for them (though add that such taboos are sometimes broken in secret at times of drought); the neighbouring Kikuyu have less close working relationships with donkeys, and do eat donkeys. Twerda *et al.* (1997:51–2) contrast the Samburu, who 'regarded it as being inferior to consume donkey products other than for medicinal reasons', with the neighbouring Turkana, who eat donkey meat and are seen as 'a hardy society used to surviving in harsh conditions where every possible source of food must be used': here, in contrast to the Native American example, equids are consumed by the poorer group.

Taboos may overtly relate to religion; in present-day developing regions, donkeys are commonly not eaten in Hindu, Buddhist, Islam, Jewish and some Christian societies, while the Confucian religion does not ban their consumption (William Clarence-Smith pers. comm. 2016). There are nuances in this analysis, with Muslims seemingly forbidden to eat domesticated donkeys but not wild ones, and the pattern of modern donkey-eating generally features criss-crossing of religious, cultural and lifestyle factors. Blench *et al.* (2004:213) suggest that outside the donkey's native habitat 'the donkey is an exotic to which no culinary taboos attach', but there seems an economic objection to this hypothesis. Taboos may make donkeys less attractive to thieves than cattle but can lessen the perceived desirability of donkeys over working cattle, as donkeys are not useful for subsequent selling for meat.

Hides

Donkey hides are tough and elastic so have a range of uses including for parchment, but unlike cattle hides, they are not widely used in modern

Africa, perhaps through association with the taboo flesh, though donkey tail-hair is sometimes used and there is a massively increasing export trade in hides to China for use in medicine and cosmetics. The wild onagers indigenous to the Near East may well have been hunted for hides, as well as meat, from an early date (for example at Umm Dabaghiyah), and Zeder (1986:404) has suggested that donkeys may have been among the equids butchered for hides at 3rd-millennium BC Tal-e Malyan in SW Iran, from the equid bone distribution. There is some Ancient Near Eastern textual evidence of skinning for hides in the 3rd millennium BC (e.g. Adams 2010:5), and at 3rd-millennium BC Khirbet al-Batrawy and 2nd-millennium BC Tell Jemmeh donkey remains show some cutmarks indicating skinning for hide (Alhaique 2008:340). Overall, though, there are few archaeological indications that domesticated donkey hides (and sinews, etc.) were systematically harvested, perhaps because the supply of dead animals was not centralised. Postgate (1986:203) reports, for example, that the domesticated donkey carcass in the ash-tip at 3rd millennium BC Abu Salabikh in Iraq had been deposited complete with hide.

Research into the use of equid-fat in historical times for dressing sailcloth (a process known as smörring) offers an intriguing potential explanation for any occasional finds of fragmented donkey bones and a warning against simplistic food-focused hypotheses. A conference paper by Nimrod Marom *et al.* (ASWA, Barcelona 2019) concludes that such a find at the late Roman site of Magdala suggests this usage for carcasses not culturally considered edible.

Milk

Donkey milk is high in lactose and very close in composition to that of human milk; it was used for human babies in 19th/20th-century AD France (e.g. Audiot and Garnier 1995:66), and mentioned occasionally in ancient Egyptian texts, though not in Mesopotamian ones. Donkeys rarely produce excess milk, and in Africa the milking of donkeys is not widespread, because the milk is often seen as 'unclean', but also probably due to low-management systems in which milking would require regular catching of the animals; other more productive animals such as goats are, too, more readily available. Milking of donkeys (and use of blood from live animals including donkeys) is practised by a few pastoralist groups such as the Maasai and Turkana; donkey milk is also consumed in Mongolia and China, but its use elsewhere in the modern world is rare. Donkey milk is considered in some modern African and Chinese cultures to have medicinal and magical qualities, and in Greco-Roman times it was recommended for cases of poisoning and snakebite, various illnesses and aches, healing of wounds, etc., as well as being used for cheese by the Phrygians (Bodson 1985:6) – though it is low in fat and protein, so not easy to turn into cheese. Donkey milk is also used in some African and Western cultures for cosmetic purposes.

Dung

The use and virtues of dung is a large and growing subject, but outside the scope of this book. In the 20th century AD there were various unsuccessful drives by authorities in Africa and elsewhere to encourage farmers to cultivate in a new mode, using manure from working animals rather than the traditional shifting or fallow systems. A factor in the failure of the model was the consequent distancing between animal-keepers and agriculturalists, as there were no longer opportunities for pastoralists to graze their animals on resting land. Dung collection from working animals can nevertheless add to their profitability, providing fertiliser, fuel and construction material, and donkey dung is recorded as having been used in Greco-Roman times (Bodson 1985:6). Donkey dung is fibrous, which may have been valuable in use as temper for manufacture of pottery. Views on its value as fertiliser vary, and the dung is drier than that of cattle and so does not disappear into the ground so quickly (though this aids collection).

4.3 Hiring, lending and sharing

Introduction

The spreading of working-animal utilisation among more than one user is established practice, as with farm machinery today. A wide spectrum of arrangements for temporary usage of working animals is employed; these strong and self-generating levelling mechanisms enable maintenance of a minimum practical resource of working animals within a community. There is good textual evidence that this was also an active facet of ox and donkey employment by at least the early 2nd millennium BC, with arrangements between owners and other users recorded in legal and commercial texts from Mesopotamia, Egypt and Anatolia.

Strong modern evidence demonstrates that, in an adjustment to traditional archaeological models of ox-ploughing adoption, the economic importance of year-round team utilisation levels is at least as much a driver for hiring out as is demand from non-owners. Insufficient attention has perhaps been paid in archaeological models to the fixed-cost burden of working ox management such as training, husbandry and foddering, and to the variety of strategies required to provide return on investment. Modern studies show that these include a typical continual shift between ownership and hiring-in of ox-teams, according to changing individual circumstances.

Communal ownership of working cattle has generally had less success in modern times than kinship group arrangements, with the latter very possibly in place in antiquity; in contrast, communal systems are often well-suited to donkeys, which are typically able to forage for themselves whereas herders are needed for leading cattle to pasture and water. Systems used in Africa and elsewhere today by both sedentary and nomadic groups, such as recruiting working donkeys *ad hoc* from a semi-wild 'pool' (see my Glossary in Appendix A), may have ancient origins and contribute to the comparative invisibility of the working donkey.

More formal hiring out of donkeys is also very commonly noted in modern African studies; typically, the recipient is using the donkey for income generation, through transporting goods for sale and carrying goods for others.

Cattle

Owning versus hiring in/borrowing of working cattle

Investment in ploughing oxen and their equipment, and in their training and maintenance, is likely always to have imposed a significant entry cost for hoe farmers, to be recouped by the presumed benefits. This poses dual questions on possible social and economic responses to the early systematic adoption of ploughing oxen: can communities develop successful levelling strategies to reduce potential ox-related wealth disparity? Can owners implement systems which provide sufficient return on investment? In this chapter I argue that on the basis of robust evidence from modern Africa, the very widespread formal or informal temporary use of ox-teams and their equipages by non-owners goes a long way towards supporting both hypotheses.

Embarking on production of a monocrop agricultural surplus ties farmers to narrow windows of activity, with rapid cultivation capabilities potentially ranking even higher than ownership of land; McCann (1995:80) argues with reference to recent Ethiopia that '[u]nless households owned oxen or had access to them, land had little economic meaning'. As is often argued in support of inequality deriving from ox-ploughing, those without working animals have to hand-cultivate, hire in or share animals, leading to late ploughing and, so, to late sowing, with resulting increased risk of low yields or outright crop failure. Non-owning households may also have to lend household members for labour in exchange at key times at the expense of their own fields.

The actuality is often more flexible. There is modern analogical support for a model in which the burdens of hiring or borrowing ploughing animals form an incentive to or preparation for ownership. The pattern of animal ownership may shift frequently within a community, with farmers switching between owning and hiring according to their situation and to external factors such as changes in maintenance costs or income. In Nigeria, according to an Overseas Development Institute overview of West Africa traction-animal use,

> [f]ew farmers can afford to operate a plough without hiring it out, as the cattle have to be provided with *dusa* (bran) and *harawa* (dried fodder) during the farming season. . . . in a sample village in Kano State, ploughs were owned by only one family in 25, but . . . virtually all families could manage to hire a plough for basic ridging. (Blench 1997:15)

The study adds that hiring in of cattle may delay ploughing but can be well-compensated for by the freedom from working animal management costs. In a district of Zambia Kakwaba and van Leeuwen (1999:308) similarly report that

> [o]nly 13% of rural households own oxen but approximately half the cultivated area in Kaoma District is plowed under hiring and borrowing arrangements with ox owners. Most (51%) of oxen exchanges were on a sharing basis. . . . More than half of hiring transactions were to relatives, and relatives, especially women, had a high priority in terms of time of plowing.

Owners may lend young cattle to others for grazing until they are old enough to train, with owners then ploughing the grazing-partners' fields as well as their own (Mgabo and Masune 2012:192, on Tanzania). Pingali *et al.* (1987:112) conclude from studies that a strong self-regulating factor within African communities, often in the form of hiring or lending ox-teams, soon erodes any competitive advantage from ownership of working animals. Other writers on Africa reinforce the speedy reversion to equitability that commonly occurs. A balance can also be made between human and animal labour, depending on the cost of hiring: for example, in Mali in a survey in the 1970s (Delgado and McIntire 1982:190–4), farmers found it very profitable to substitute rented oxen hours for hired human labour hours in the July seedbed preparation bottleneck, as long as the hired labour remained available for harvest work later.

Collective/cooperative/communal ownership

Hardin (1968:1244) was the initiator of the model of the tragedy of the commons, in which short-term self-interest leads to the collapse of communal stewardship systems such as grazing facilities and herding; but Dietz *et al.* (2003:1907) point out the many examples where 'self-governing institutions' have prevented, or at least delayed, this. Ostrom *et al.* (1999:278) use examples in north-eastern Asia to demonstrate that Hardin's model is a worst-case one and less likely to apply to an established cultural group and to a situation where (p. 281) 'the flow of resources is relatively predictable'. There has been only limited research into the present-day relationship between cooperative/communal activity and animal traction (Farnham

1997:30); the few examples available provide valuable pointers for potential detection in antiquity.

Gboku (1988:317) reports on studies of working-animal adoption in India and Nigeria, in which high-adoption villages displayed a distinctive range of indicators of an established community structure and a culture of cooperation. Starkey and Apetofia (1986:15) find contrasting results in Asia and Africa:

> [i]n both Nepal and Indonesia, it is common for villages to set aside specific areas for grazing. . . . While such village cooperation might seem unremarkable, there have been many cases in West Africa where the lack of agreement on grazing arrangements has inhibited the adoption of draft animals or restricted the integration of crop and livestock systems.

Peters (1986:141) reports on a successful system of mutual obligation and assistance in Botswana, where 'supra-household networks' arrange matters, such as supply of draught animals or of an individual to care for cattle, whilst the owner goes away to work. Isaakidou (2008:105) suggests that such 'mutual assistance or exchange' systems may be detectable archaeologically through the stylistic uniformity of ceramics resulting from a 'strong collective identity'.

While communal operation may operate seamlessly and unconsciously among kin and small community groups, there is plentiful evidence in Africa in particular that communal ownership on a larger scale cannot easily be imposed without the presence of solid mechanisms for its effective regulation. Starkey (1991:83) reports that well-meaning post-colonial agencies in several countries in 20th-century AD Africa attempted to introduce animal traction through communal ownership, with often unsatisfactory results. Key sticking points were the seasonal nature of draught animal use, resulting in demand bottlenecks and the sharing of responsibility for maintaining the animals year-round, including ensuring sufficient grazing, preventing theft and restraining them from damaging crops. This raises the question of the means of enforcement of systems such as that described in the 2nd-millennium BC Mari texts, where traction oxen and associated ploughing equipment are recorded as being government-owned and contracted out to villages, with payment in crops or manpower (Matthews 1978:86–94). There is a suggestion that this might usefully allow semi-nomadic groups to plough their land using oxen, but also evidence in the texts that many of these groups then failed to show up when summoned for labour in return.

Lending and hiring

Lending and hiring differ from communal ownership in that they benefit individuals possibly at the expense of others in the community. Issues to address here are: Does both providing and receiving temporary use of a working animal provide social and financial benefit? Is this benefit greater if a kinship-based lending system is in place rather than formal hiring arrangements?

In Ethiopia where ox-ploughing has (uniquely in sub-Saharan Africa) been practised since the early 1st millennium BC, McCann (1984, 1995) concludes through detailed historical and modern-day studies that although land was an important inheritance, the key factor in production was ploughing oxen, over both land and labour; in this pre-monetary society, intricate oxen supply and use systems evolved and continue today. Wealthy households 'rent' oxen to poorer neighbours in return for labour and seed, tying indigent households to powerful ones and promoting stratification in the rural economy. In contrast, modest households with an ox-team loan it to indigent friends or relatives, providing a levelling process at village level; and farmers with a single ox join up with another to allow ox-team ploughing for both (McCann 1984:8–9). Similar systems – known as *mafisa*, *magazo*, *kimanangan* – exist throughout sub-Saharan Africa, with cattle loaned on a basis that may be equitable or patronage-based (McCann 1984:8–9; Shenk *et al.* 2010:72). Among camel-herding cultures, too, impoverished relatives are lent camels, with the advantage of shifting the burden of maintenance and protection. There is a good economic rationale to long-term lending of cattle (Bogucki 1993:499), such as keeping up relationships, dealing with labour shortages and distributing the herd to buffer against risks such as theft, disease and grazing shortages (Russell 1998:45).

There is inevitably a blurring between lending and hiring at this point, as the borrower has incurred an obligation which might, in the future, involve supplying labour; Farnham (1997:102–4) reports that in north Togo most ox-owners had at some point lost animals and borrowed from neighbours, at no charge but with the knowledge that a favour would be returned: such lending of oxen for free was explicitly seen as 'an insurance measure' (Farnham 1997:110). Inequality may emerge temporarily in the early stages of adoption, as described by Vall *et al.* (2002:125) in a village in northern Cameroon where the introduction of traction animals produced a classic hierarchy of wealth-accumulating owners, renters unable to earn sufficient to buy because of a shortage of available land, and young hired labourers aiming to climb the ladder; the authors then contrast this to areas in Cameroon with longer established animal traction, where an equitable system of lending between kin has evolved.

As I underline, in modern Africa modern working animal overviews and studies indicate that adoption of ploughing oxen almost invariably involves hiring out the team. Tiffen (1976:128–31) describes insensitive constraints (now abolished) in a region of Nigeria in the mid-20th century AD, where farmers were supplied by central authorities with oxen and ploughs but forbidden to hire them out 'lest other farmers should be enabled to cultivate more land without using manure' (p. 128); there was therefore disappointing uptake as some farmers regarded hiring out of the plough and team as a crucial factor in acquiring these assets, providing 'a means of obtaining cash to finance the extra weeding that ploughing a larger area of land made necessary' (p. 130). McCown *et al.* (1979:305) report that some owners in Africa even 'preferred to hire out their animals and cultivate their own farmland with hired labour', and there are accounts (e.g. Lawrence and Pearson 2002:109) of landless individuals making a living simply from hiring out draught animals. Ox-owners may also hire themselves out with their team, adding to the payback on their animals; in regions where it is not considered suitable for women to handle oxen – so that they have to buy the owner's handling services in addition unless they have suitable in-house male help – this system is of particular use.

A spectrum of inequality

I discuss later, on the subject of ploughing and of the role of women (Chapter 7.3), the sometimes simplistic assumptions by anthropologists of the relationship between working cattle ownership and social inequality. I indicate in Table 4.1 the spectrum of practical arrangements in place in modern developing regions with significant use of working cattle today, as an indication of the dangers of binary thinking in considerations of practises in antiquity.

Payment

How is the 'cost' of providing working cattle to others calculated? In Mesopotamia, with its centrally owned ploughing animals, there are few textual references to cattle hiring or sharing until late in the 3rd millennium BC, when for example in a Girsu text, rent for plough animals is recorded in litres of barley (Heimpel 1995:88). In early 2nd-millennium BC texts, Stol (1995:185/191/198–9) records regular references to cattle being hired for threshing/ploughing/pulling a wagon, for barley or shekels, including indications that the rental fee was on occasion shared between two people, with co-ownership of an ox. He makes the point (1995:198) that human members of the plough team are similarly hired, underlining the seasonal nature of ploughing and the occupation elsewhere of humans and animals for the rest of the year; this also adds dimension to the exchange of use of working animals for human labour, common throughout history to today.

At family or small-community scale in Africa, there is a blurring of the boundary between lending and hiring of oxen. If a more formal recompense for the use of ox teams (with or without the owner) in Africa is to be made, it may take the form of labour, cash or produce; payment may be deferred until after the harvest. Reports are given of between one and five or more person-days exchanged for an ox-team day. Peters (1986:150) notes that in Botswana, outside close kin, only male labour can be exchanged for ox-team

Table 4.1 Levels of potential social inequality from ploughing-oxen ownership arrangements

System	*Operation*	*Potential social inequality*
Individual ownership	Use never open to many (unless subsidised)	High inequality
Individual ownership plus commercial hiring	Allows others with some means to use draught oxen, but is not open to all and can exacerbate inequality if entrepreneurs drive up the price of hiring	Moderate inequality
Ownership plus a system of mutual obligation and assistance, though with specific and formalised recompense for the loan, e.g. labour required at key times	Open to most but does not materially reduce inequality (Isaakidou 2008:105)	Moderate inequality
Ownership plus a system of mutual obligation and assistance with long-term benefits	Used widely in Africa among kin and neighbours; is open to all	Likely to be the point on the spectrum where there is least social inequality
Communal or cooperative ownership of oxen	Can be open to most but has been found in many cases in Africa to be unsuccessful as a system	Low inequality in theory but not in practice
Central ownership, and hiring out or lending, as indicated for example in the Mari texts	Open to all who submit to central control but therefore imposes a different sort of inequality from one of simple financial exclusion	Inequality of power rather than finance

hire, presenting female small-scale farmers with a problem as their labour is normally only exchangeable for cash, goods or a share of the harvest.

Although there is little mention in modern accounts of compensation in case of injury to the animals, several 2nd-millennium BC Sumerian and Old Babylonian texts (Roth 1980) demonstrate that laws were in place from ancient times to protect both the hirer and the owner in the event of harm to hired oxen, concerning the degree of liability in various circumstances and the payment for death or injury claims; the Code of Hammurabi also specifies a higher hiring fee for the two front oxen in a multi-pair team, perhaps because they are more highly trained.

Donkeys

As I discuss in other chapters, there is ample evidence that the different physiological and behavioural characteristics of donkeys and oxen lead to different usage and ownership patterns. It can be suggested that donkeys are easier than oxen to own and to employ all year round, without the incentive to hire or lend, but there is nevertheless good evidence discussed later of a flourishing system today of hiring, lending and communal ownership of donkeys, including 'contract' use for carrying others' produce and goods. Texts from antiquity also support the hypothesis that the hiring, lending and sharing of donkeys was a central part of the donkey-using industry, enabling wide usage and allowing wealth disparities to be bridged.

As with oxen, both immediate benefit (e.g. grazing and protection from predators) and long-term social benefit can be achieved by lending donkeys. Donkeys are lent to help relatives and friends or to establish good relationships with neighbours and local groups (e.g. the large Admassu and Shiferaw 2011 survey in Ethiopia – p. 8); Pearson *et al.* (2001:23) report from an Ethiopian study that, as a result, 'donkey use seems to be part of the social network' in rural areas. Waithanji (2009:34) reports as a practical point from a survey in Kenya that such lending 'is free to discourage further borrowing as paying for the donkey gives the person renting it a sense of entitlement to the donkey'.

Accounts from sub-Saharan Africa underline the common hiring out of donkeys as a valuable source of income, in cash or, for example, in a percentage of the product carried; while the landless without donkeys may hire one in for conducting income-earning activities. In a particular large-scale Ethiopian example, Zenebe and Fekade (2004:71) describe donkey-hiring at the huge Yehil Berenda grain market in Addis Ababa, where around 3,000 donkeys are hired out like taxis for transporting the grain purchased at the market (Figure 4.5):

> the donkey operators bring all of their donkeys to the marketplace as it is difficult to arrange for herding around their residences. Moreover, they do not want to forego business if they happen to get a lot of grain to transport. They can also rent their donkeys to other donkey operators.

The hiring out of caravan animals is mentioned in the 2nd-millennium BC Kaneš texts concerning the pack trade between Anatolia and Aššur; new animals, too, were required *en route* when donkeys fell ill, died, were stolen or strayed, and Dercksen (2004:261–3) suggests that there may have been established local businesses hiring out donkeys for a stretch of the route, sometimes with a driver, and reclaiming the animals at the next town.

A text from 2nd-millennium BC Deir el-Medina in Egypt suggests donkeys being hired out *ad hoc* for a few days, with the owner perhaps taking the donkey back home for the night (Janssen 2005:11). Janssen (2005:110)

Figure 4.5 'Taxi' donkeys waiting for hire at the Yehil Berenda grain market, Addis Ababa, Ethiopia

Source: © JG Ethiopia 2014

and Janssen *et al.* (2003:26–7/44) report on a puzzle in several Deir el-Medina texts in which woodcutters (and also water-carriers, doorkeepers, policemen, a fisherman and a potter) hire donkeys from workmen, sometimes for only a few days and at high rates. Mitchell (2018:50) explains this as indicating that donkeys were too expensive to be owned by such low-class workers, but the earlier commentators' suggestion is that these recorded instances are in fact exceptions: woodcutters and water-carriers have daily need of transport and donkeys were probably normally obtained from elsewhere, perhaps from a communal pool; this is a very useful reminder that texts might record exceptions rather than the norm. Communal ownership of a 'pool' of donkeys might be more successful than communal oxen, as donkeys can often be left to forage for themselves and are less subject to theft, as they are less in demand than oxen for meat; need, too, is generally more year-round. As I note in Chapter 5.1 on the herding of donkeys, pastoralists in Africa habitually recruit donkeys *ad hoc* from semi-wild groups when moving camp, a practice which may even reflect original domestication processes. However, Blench *et al.* (2004:217) give a vivid account of the demise of communal donkey 'ownership' in Nigeria:

> [i]n northern Borno State it was said that, until the 1950s, donkeys were semi feral and were allowed to graze freely by rivers. Individuals

who needed pack animals captured the wild donkeys, trained them, and sent them back to the river when they were not needed. As the taste for donkey meat developed thieves came from outside and loaded all the feral donkeys into trailers and sent them for slaughter. This created a shortage, and farmers who previously had considered donkeys a free resource were now obliged to buy them and to tether them in order to discourage thieves.

5 Training, husbandry and feeding

5.1 Working-animal training and husbandry

Introduction

The modern development studies in regions of sub-Saharan Africa afford a particularly rich opportunity to study eye-witness accounts of adoption of working donkeys and cattle by cultures formerly using hand-cultivation and human porterage, bringing to life the sometimes sterile references in archaeology to diffusion and establishment of the new technology. In this chapter I examine the central role of person-to-person dissemination of the concept and best means of use of working donkeys and cattle, and describe the varying approaches to working-animal training.

I explore the animal-choice implications of the contrast between the low-level management systems typical for donkeys and the intensive herding and maintenance generally necessary for working cattle. Corralling of cattle may relate simply to fear of theft, but evidence of corralling of donkeys – for example through archaeological detection of dung concentrations – can shed valuable light on how and where they were used for work.

I include here a brief discussion of the issue of castration, often cited as an indicator of working cattle, though the reality is more complex. Finally, a note about harness, though this subject is well-covered by other authors.

Aspects of training of animals and users

The role of person-to-person dissemination

Archaeological commentators tracing the diffusion of working animals across space and time tend to focus on the movement of the animals and associated technology, rather than of instructions for use. There has been almost no envisaging of the actual process of adoption of working animals at ground level, particularly in the case of donkeys (introduced domesticates in Mesopotamia); the modern studies give strong reinforcement to a person-to-person information transfer model, subverting official initiatives. African experience indicates that it is the person-to-person spread of understanding of the means of managing donkeys or cattle that precipitates widespread adoption of one or the other. In western Ethiopia in 2014 I spoke, for example, to a farmer in a transmigrant community who had come from a cattle-using region, but on arrival had seen a neighbour ploughing with donkeys, so had gone to him to learn how to use them and to make the harness. Mutua (2004:101) reports that in Kenya among farmers accustomed to cattle-ploughing,

> [a] recent demonstration on the use of donkeys for tillage purposes in Machakos was received with disbelief. It turned out that farmers in the area had never seen a donkey plowing. They could not believe their eyes when this was done on their own farms using a donkey borrowed from a farmer who thought it could only be used for carrying goods on its back.

Similarly I met elderly farmers in central Burkina Faso in 2013 who explained that before the 1960s they had kept cattle and were aware of pack-caravan donkeys but had had no thought of using animals for cultivation until the concept was introduced by French post-colonial organisations; in the south-west, farmers had commonly held back from ploughing-animal adoption until young men returning from emigration brought eye-witness accounts and the funds to invest (and of course preconceptions as to which species to choose). Resettled groups and returning wage-migrants in several parts of sub-Saharan Africa are reported by regional experts at ATNESA workshops as having provided the concept and impetus for adopting animals for work (e.g. Sosovele 1994:318–9; Starkey 1994a:78); once implanted, new practices diffused from village to village, as with the southward spread of donkey use in regions of Africa (Starkey 1994b:1). Starkey (1992:21) also reports that oxen were promoted for ploughing and transport in several West African countries in the 1970s–80s AD, but that the farmers became more successfully introduced to working animal adoption by their close cultural links with neighbouring regions such as Senegal, where donkeys were the established work animal. For ploughing and other agricultural work in particular, farmer-to-farmer training and advice is widely reported in NGO and official studies as the most effective adoption incentive:

> Enthusiasm and knowledge is best transmitted farmer to farmer; it has been shown that draft animal technology can travel quickly without government intervention, even across borders as long as there is the need and contact. (Tibbs 1989:12, on China)

Sosovele (1994:319) comments, though, that this may disenfranchise women as male farmers tend to teach males; in Ethiopia, Pearson *et al.* (2001:30) report that women often ask male neighbours or relatives to train their donkeys for them.

User induction issues are exacerbated in regions without a tradition of keeping livestock, though Gboku (1988:315) and others report from Sierra Leone that even in cultures already familiar with the animals, a positive interest in using them for work – not necessarily present among traditional cattle herders, or correlated with farmer age – is equally important. Ignorance about proper use of animals leads to reports of 'fear of cattle' (Reh 1982:77) as one of the most-mentioned factors in non-adoption (Table 7.1 on uptake factors), and to inefficient use resulting in abandonment due to lack of work benefit. New users might therefore prefer ready-trained animals initially; hiring of working animals before investing in them is often a valuable process. Modern African studies mention specialist ox-trainers in villages, perhaps implanted by the authorities.

The introduction of working-animal use to new adopters not only includes handling and maintenance of the animals and implements, but also (for ploughing and other field tasks) in devising a revised crop strategy and agricultural schedule to make the best of the new technology. The user training process in the 2nd millennium BC is clearly illustrated by the Sumerian text *Farmer's Instructions* (also known as the *Farmer's Almanac*), which describes proper cultivation of cereal crops with working animals over an agricultural year, for a farmer with some basic knowledge but in need of advice on best methods (Black *et al.* 1998–2006d; Maekawa 1984). There is ample evidence in African studies and elsewhere that lack of such farmer training leads to low adoption or abandonment of working animals, particularly for ploughing; and similarly, that productivity grows with experience. Poor or unsuitable training of farmers (as well as of the animals) may well be a factor in the striking regional variations in the number of both humans and animals considered appropriate to form the plough team, as I discuss in detail in Chapter 7.2.

Training of animals

Training of working animals can be crucial to effective work output, in terms of speed and accuracy of work, and also importantly in the number of handlers required. Training of animals also increases their value as an asset, though brings its own constraints: animals, especially cattle, may forget their training if not in frequent human contact and not kept separate from the herd (see earlier on herding of working animals). A study by de Wilde (1967:112) reports that often in sub-Saharan Africa '[i]nadequate training has resulted in poor cultivation and excessive employment of labor', with several people employed to handle a single-ox plough.

> One of the principal difficulties is that the animals do not stay trained because they are normally harnessed only for the brief plowing season. They would become more tractable and efficient if they were used more continuously.

Trained animals may be less likely to stray, but trained cattle in particular are easier and more desirable to steal. The investment, too, needs to be literally 'sweated' year-round, particularly if main usage is seasonal, for return on investment and to avoid time-consuming retraining.

Training of donkeys for pack is generally straightforward, as tamed donkeys soon accept loads and are behaviourally suited to following in line and learning routes; in modern donkey caravans, the animals are generally not tied but follow freely. In 3rd-millennium BC Egyptian depictions, foals are often shown accompanying work donkeys, as they learn in this way from their mothers (Power 2004:136). Donkeys in modern Africa are sometimes recruited from pack-trains for ploughing, and Mutemi (2008:69), a donkey merchant in modern Kenya, reports that donkeys found unsuitable for ploughing are resold for pack.

Training for the plough includes teaching animals to walk in straight lines and recognise furrows, to turn neatly when needed, and ideally to respond to voice commands, which reduces the number of handlers needed (Figure 5.1).

> In both Nepal and Indonesia, working animals are invariably controlled by just one person. . . . In Nepal, animals are often used without nose ropes or reins . . ., and the basis for control is verbal commands and the tap of a stick when necessary. (Starkey and Apetofia 1986:16)

If the animals are used in teams, there is also the task of teaching them to work together. The 3rd- and early 2nd-millennium BC Sumerian and Akkadian texts on ploughing distinguish between different oxen in a team, probably in relation to their different type and extent of training; untrained

Figure 5.1 Boy training cattle for ploughing, Jawi, Western Ethiopia, familiarising them with the whistled commands and lifting and lowering the ard to accustom the animals to the task

Source: © JG Ethiopia 2014

oxen are assigned particular places in the plough team, and oxen trained for specific tasks such as ploughing, seeder-ploughing, threshing and non-plough draught (Renger 1990:275–6; Stol 1995:184). The 2nd-millennium BC Babylonian Code of Hammurabi specifies different rents for the various oxen in a plough team, while the contemporary Sumerian 'Laws about Rented Oxen' specify significant compensation if a ploughing ox is injured (Roth 1980:127).

Training strategies

Donkeys are widely acknowledged in modern Africa to be easier to handle than cattle, and behaviourally more susceptible to cooperating with humans and understanding skills such as ploughing, though are famously stubborn if they consider the task unreasonable. A strategy recorded by Schmitz *et al.* (1991), suitable if plenty of household or other unpaid labour is freely available, is to skimp the training and only use the animals for a short period per annum, using several people to control the team. In North Africa the Tuareg rely on rough treatment to 'tame' donkeys for riding and pack (Nicolaisen 1963:109), despite widespread evidence (see earlier) that this is not necessary; Hagmann and Prasad (1995:235) and Waithanji (2009:19) report that donkeys in Zimbabwe and Kenya respectively are not trained but beaten to achieve the work needed. Such strategies may be viewed as necessary where farmers lack the knowledge to train animals to a point where a single user can handle them. In central Burkina Faso (JG Burkina Faso 2013) I observed widely differing strategies concerning plough animals within a small district depending on circumstances; two young men were pushing and urging an untrained pack donkey up and down to plough, while in contrast a few fields away, a young woman (who ploughed alone as her husband worked elsewhere) had trained her donkey to respond to voice commands and to turn without prompting (Figure 5.2); in contrast again, a farmer in the adjoining field bought young bulls from a nearby abattoir, trained them for a few days and employed them for ploughing (requiring several family members to cajole and steer the animal) before returning them to the abattoir (Figure 5.3).

Herding of working animals

Free-living donkeys

The many African and other accounts of semi-wild donkeys rounded up *ad hoc* or seasonally for work (e.g. Kimura *et al.* 2010:6, and see the Glossary in Appendix A) invite comparison with the rounding-up for work in many regions of reindeer, camels and horses. Groves (1974:159) reports on

> donkeys in the Sahara that roam freely but are owned by different people who, when they move from place to place, re-capture them, load them up and take them along – then release them at the next stopping-place.

Figure 5.2 Woman ploughing with her well-trained donkey in Ziniaré, Burkina Faso

Source: © JG Burkina Faso 2013

Figure 5.3 Family members ploughing with a young untrained bull in Ziniaré, Burkina Faso

Source: © JG Burkina Faso 2013

Such a commensal relationship is also demonstrated in the Turkana region of Kenya, where groups of donkeys live freely but are familiar over time with the pastoralists with whom they share a range, as representing an occasional source of salt, well water and protection in exchange for cooperation during camp moves; the pastoralists recognise individual cooperative donkeys, and have sophisticated skills in finding and getting close to them (Stephen Blakeway pers. comm. 2017).

I describe in detail in Chapter 5.2 how working donkeys, who are well-equipped to deal with predators and find food and water, are still today often left to graze unguarded, in mixed natural groups outside the settlements, or herded on verges and communal land or with other livestock. It is a sad footnote that in Africa today they are increasingly having to be corralled or kept near settlements at night as demand for their donkey hides for export to China (and so theft) rises.

This leaves the conundrum of large donkey herds reported in antiquity. An Egyptian text from the mid-3rd millennium BC refers for example to an individual owning more than 760 donkeys (Osborn and Osbornová 1998:135), but these were very unlikely to be kept as large captive herds: donkeys naturally live in small groups rather than large herds, and are capable of returning to base unaccompanied, and their valuable ability to roam widely and find rough food for themselves would be severely restricted if kept captive in large groups. A suggested explanation is that most of the donkeys concerned were out working, in caravans for long periods or in the temporary care of individuals using them for a range of agricultural and transport work, rather as an owner of a fleet of rental vehicles nowadays does not expect to be housing many of them at any one time.

Herding of working cattle

Oxen are considered to require a good deal of attention from their owners (e.g. Sosovele 2004:107), in both care and general supervision, including the sometimes elaborate measures necessary, as they need long periods in daytime to graze and ruminate (unlike donkeys) and daily access to water. Several accounts of working oxen in Africa refer to their unfriendly and sometimes aggressive nature, an element sometimes cited as constraining women from working with oxen although cattle are habitually handled and herded by women and children. The suggestion of using cattle from semi-wild herds for work, as with donkeys, is never mooted by commentators; Ingold (1980:140) notes that 'untamed cattle could only be managed on a ranch basis; ranch cattle are agile and fierce', though Groves (1974:159) reports that 'in Assam there are cattle that live free in the forest, but come to the villages to be fed, milked and slaughtered' (but not worked). Cattle temperament can be improved by frequent contact with humans from when young, and by working regularly with the same individuals (Reh 1982:77/108; Renger 1990:271; Schmitz *et al.* 1991:79; see what follows on training of working cattle); for this reason, female cows, with their regular human contact during milking, are typically more tractable.

Working-animal herders

Pollock (1999:103), examining gender in late 4th-millennium BC depictions of humans tending animals, suggests that it appears to be men who feed animals, individuals displaying no gender indications who master animals, and mainly women or genderless individuals who carry out other activities involving animals. Zarins (2014:230–1) records references to apparent professional donkey herders in the late 4th millennium BC (in Fara and possibly Uruk Archaic Texts) and mid- to late 3rd millennium BC (late EDIII/Old Akkadian texts from Ebla, and an Ur III text (BM 19984) concerning rations for an overseer of equids), with (p. 238) a table listing 'Personnel Associated with the Girsu Equids in the ED IIIb Period'; he remarks (p. 256) on the increasing mention from the mid 3rd millennium BC of stables and compounds for equids. Landais and Lhoste (1990:224) report that little work has been done in modern Africa on evaluation of the labour required for animal husbandry, especially in the case of cattle, but that the value added can be high, particularly as animal feeding and herding work is often carried out by quite young children or by indigent family members. McCann (1984:4) reports from West African study results that the additional labour needs for working oxen 'are distributed throughout the year and can be absorbed by younger members of the household (average age of livestock caretakers in the West Africa case was 12.6 years)'. In a study in Burkina Faso reported on by Singh (1988:31), 50% of the children's household and farming labour time was spent this way, with older male children in particular concerned with animal traction activities.

Hobbling of animals also reduces herding burdens where crop protection and straying are more of an issue than predation and theft.

Animal theft and corralling

Late 3rd-millennium BC texts from Girsu and the 2nd-millennium BC Archives Royales de Mari (ARMT II 123:15–24) show that cattle raiding was a significant activity (Heimpel 1995:106; Matthews 1978:104–5); Porter (2012:22–3) discusses likely political causes, but Matthews (1978:125) notes that the raiding system has practical and social effects, as raiding reduces the tendency for 'clumping' of large herds owned by the very rich, evening out social inequality and also reducing the effect of disasters such as drought. The problem of animal theft is not confined to working animals, but as farmers point out, trained animals are more docile and easier to steal, and the lost investment is greater. Before the time of horses for riding (which allowed animals to be driven rapidly away), large-scale theft may have been more likely in populous areas where issues of concealing stolen animals are fewer (Russell 1998:49) and more disposal outlets are available. Urban populations, too, may generate a high demand for meat (Schmitz *et al.* 1991:80). A modern urban issue is the killing of donkeys in South Africa, apparently for reasons of competition between carters; in rural areas this manifests itself as theft or unauthorised use of donkeys.

High incidence of theft – or the fear of it – can deter farmers from using working animals at all. It may also affect choice of working animal in favour of donkeys over cattle, as the former have lower economic and social value and are not eaten in many cultures so theft incidence is lower; trained cattle are, too, a valuable asset, requiring as they do more training than donkeys (see the examples to follow). There are multiple reasons as well as fear of theft for keeping valued animals in pens or corrals, or within the village confines: fear of straying, predators, protection from heat, cold or damp, protection of growing crops, collection of dung, intensive training, convenience of keeping working animals nearby. Corralling animals raises feeding and watering issues, as fodder needs to be carried, or the animals have to be taken long distances to drink and graze; evidence of corrals may therefore indicate an intense level of raiding/theft.

Little attention has been paid until recently to the archaeological detection of corrals and other arrangements for restraining animals used for work. McShane and Tarr (2007:125) point out that even in 19th-century AD United States written history, '[s]tables rarely make it into histories of the built environment, although they constituted a substantial part of the environment'. Structures for keeping animals may often be temporary, making detection tricky; even sedentary groups shift their animal pens at times to fend off insect infestation and diseases, while corrals for donkeys are likely to be small as they are not normally kept in large numbers. Existing trees or poles are universally employed for tethering of animals; correlation of postholes with animal-tethering is very problematic, though Oates and Oates (2001:41) report that in the famous building in 3rd-millennium BC Tell Brak in northern Mesopotamia, where several complete donkey skeletons were found, '[t]he presence of live donkeys in the courtyard is also attested . . . by the presence of herbivore dung and the outline of stakes still used to tether donkeys at the present day'. More clues may be obtained in treeless regions: possible tethering-holes for animals (used to this day for tying donkeys) have been identified in a cliff-face at a 3rd-millennium BC mining site in Egypt (Kuhlmann 2002:133), and Brewer *et al.* (1994:86) report that tethering stones are used to identify Dynastic Egyptian cattle pens. Donkeys are sometimes simply hobbled, though they are said to have more of a tendency than for example camels and horses to remain nearby. This practice is difficult to detect archaeologically, though exostoses (bony lumps) on limb bones can indicate regular hobbling.

Palmer *et al.* (2007:369–79), in an ethnographic study of Bedouin groups in the Wadi Faynan in southern Jordan, record the archaeologically inconspicuous nature of animal-management arrangements: trees for tethering donkeys, caves for penning animals (as in Figure 5.4), predator-deterrence through flapping fabric tied to sticks. An experimental process was carried out in the 1970s in East Africa where an abandoned pastoralist camp, last inhabited 10–20 years before, was investigated for archaeological evidence of a pastoralist lifestyle (Robertshaw 1978:29). No postholes, for human or animal habitations, were found, but dung was evident in what was seen from the surviving fence to be an animal corral. At the time, it was concluded that no detectable evidence of animal-keeping would survive at prehistoric sites, but more recent technologies in micro-analysis of sediments make it possible to identify elements of dung remains indicative of species and diet as well as of penning and annual cycle of on-site presence (see Redding's example earlier).

Unlike cattle dung, the presence of donkey dung accumulations (outside their natural habitat) is a significant clue to their working use; detection of dung deposits is gaining ground as an indicator of pastoralist and other enclosing of animals.

Donkeys enclosed at night by pastoralists and farmers are sometimes kept in specific enclosures away from cattle, to avoid horning incidents, with consequent distinct dung accumulations.

> [M]icromorphology, phytolith concentrations, mineral assemblages and the isotopic composition of organic nitrogen could be used to detect the presence of ancient donkeys [sic] dung following approaches developed by Shahack-Gross and colleagues (Shahack-Gross *et al.* 2003, 2008) to identify cattle, sheep and goat dung in degraded sediments in

Figure 5.4 Rural animal pen, NW Oman

Source: © Jill Goulder 2013

> prehistoric African pastoral sites. Because equids are not ruminants, larger fragments of vegetation make donkey dung readily distinguishable from that of cattle, sheep and goat. Pellet size and shape, hoof conformation and trampling morphology should also distinguish equid from bovid accumulations. (Marshall and Weissbrod 2009:75).

Shahack-Gross *et al.* (2008:984) note though that dung from browsing animals (and therefore donkeys) features fewer phytoliths than that from ruminants.

Detection of these keeping-areas could be impeded by the common practice throughout history of collecting dung into a single heap near to dwellings; but area analysis might nevertheless detect dung-containing areas devoid of occupational debris, or evidence of arrangements for drinking-places, and at a higher level Corona satellite imagery of 'hollow ways' from the 1960s might pinpoint animal-route nodes, both local from ploughed fields and longer-distance.

Castration

Osteological evidence of castration (seen for example in greater slenderness of metapodials) is one of the potential indicators of the early use of male cattle for work, though the correlation is far from exact – and of course does not aid in detection of working cows. Castration of bulls is widely known to produce a quieter temperament, but in modern developing regions quite commonly working bulls are not castrated; reasons for refraining from castration of working bulls range from a need for males for reproduction, the greater strength and resilience in difficult conditions, use of a docile breed, or simply lack of good knowledge of the castration process and benefits (e.g. Blench 1997:15). It is, too, the strongest males who are selected for work so 'the breeding bulls may be genetically inferior in body size and conformation' (Starkey 1991:78). Gaastra *et al.* (2018:1464) suggest that animal traction, since earliest adoption, has included *ad hoc* use not easily detected:

> The occasional use of a good-natured cow or young bull to haul timber would be considered traction exploitation, but one that would not necessarily alter management profiles or significantly affect bone development across an entire faunal sample.

Evidence of castration in the past may indicate more focus upon ploughing animals for arable than raising for beef, or perhaps upon animals intended for specialised work requiring good control, such as hiring out for work or complex tasks such as forestry, and pack and riding.

Incidence of castration of working donkeys in modern societies varies by region but is not customary. Male donkeys are not generally aggressive when working, and owners may be conscious of the advantages of owning an intact male, in view of the fairly slow reproductive process. The physical consequences of castration of donkeys are much less marked than with bulls, and accounts of beneficial changes in strength and temperament are largely anecdotal. Uncastrated males may exhibit difficult behaviour in the vicinity of fertile females and may roam in search of them, affecting availability for work and ease of handling, and there are accounts of *ad hoc* castration to deal with such issues.

Maekawa (1979:102) notes that in Sumerian texts female equids are more commonly used than males for ploughing (were males instead employed further afield for pack?), and that males used were systematically castrated – perhaps because of potential issues with the largely female teams; Heimpel (1995:108), though, reports from Girsu texts that in the late 3rd millennium BC male ploughing donkeys were not necessarily castrated. I mention in Chapter 4.1 the evidence of use in 3rd-millennium BC Mesopotamia of constraint features such as genital strapping. Pearson *et al.* (2001:66) report that in Ethiopia female donkeys are used in the dry season for transport and in the wet season for breeding, and high-intensity usage in the 3rd millennium BC may have included restraints on breeding during the working season.

Uncontrolled breeding is not typically a major concern for modern individual donkey owners, but Marshall and Weissbrod (2011:S406) report on selective castration of donkeys by Maasai pastoralists to eliminate undesirable physical or behavioural traits from the breeding pool; there is the added breeding issue of the potential for reduction in size and strength if weak animals are not eliminated in this way.

Harness

The subject of harness in antiquity has been well covered by writers such as Littauer and Crouwel, with discussion of its effect on working animal adoption and performance; depictions and texts indicate use of nose-fastenings and nosebands for cattle and equids, and there is much discussion of early bit evidence. Use of yokes is indicated in 4th/3rd-millennia BC Mesopotamian representations and descriptions of harnessed cattle and equids, and commentators suggest that early adoption of donkeys for ploughing may have been constrained by transfer of the unsuitable yoke system. Yokes are designed for the physiology of cattle, with their strong vertebral arch, and are unsuited to equids, which concentrate their strength in the chest. It is possible that the representations and mentions of yoked animals reflect elite practices and that other harness types were used by small-scale farmers, as with the simple traces used from early times for pulling sledges. In many regions today simple neck straps or collars are effectively used, with equids trained to respond to voice and light touch; but even today, donkeys are sometimes yoked, through ignorance or to save the cost of separate donkey and cattle harnesses. Modern Western discussions emphasise harness efficiency rather than expediency, but there is enduring evidence from history to today that individuals and cultures do not always use the most effective harness despite the loss of efficiency.

5.2 Grazing, fodder and feeding

Introduction

Modern African studies provide valuable examples of the feeding issues presented by operation of working animals unable to graze and needing extra calories and nutrition. Despite evidence in Ancient Near Eastern texts and zooarchaeology of foddering of animals, there is still very limited archaeological discussion of the certainly sizeable foddering industry associated with early systematic use of working animals, and of its effect on labour patterns and the agricultural landscape. There is scope for new analyses of the implications of shifts away from human food crops, particularly towards those offering concentrated nutrition for animals. These fodder crops (barley, alfalfa and other hay crops, bitter vetch, foraged wild plants) may be located on land further from settlements, or in place of over-grazed pasture, leading to increased labour needs for carrying fodder and for keeping cattle penned. Intensive grazing of cattle is recorded in antiquity and has further potential for archaeological detection as evidenced in modern examples, which indicate that the presence of such practices has particular implications for social renegotiation in areas such as communal grazing.

Meanwhile, robust modern evidence demonstrates that key differences in maintenance needs between donkeys and cattle lead to very different trajectories as work animals. The extra feeding required by working cattle can provide eventual payback in terms of carcass value, but there is a marked shift among farmers in Africa to the lower-investment strategy of donkey power. Donkeys are physiologically more efficient than cattle at food and water processing and more behaviourally flexible in their feeding needs. Unless they are far from natural forage resources and engaged in heavy pack or vehicle work, they are often self-maintaining. Particularly in times of environmental stress, their ability to work without concentrated pre-feeding in the dry season has contributed significantly to their increasing popularity over working oxen in Africa.

Accounts from throughout the African continent, of the expectations and actualities of adoption of working animals by farmers, give invaluable pointers to the learning curve experienced during their early systematic use. The modern evidence is that the drawbacks of adopting cattle for work (notably labour demands for foddering and penning) may have been outweighed by levelling factors such as hiring out. Working female cows may be favoured over oxen in regions 'characterized by insufficient feed resources to justify maintaining non-reproductive animals' (Starkey 2011:25). A different equation applies for the adoption of the donkey despite its more limited meat and milk potential: the modern evidence worldwide indicates the high work return on very low foddering and grazing outlay.

Fodder and nutrients

Extra feeding of working animals

When working animals first replaced humans, a significant element of the intended economic benefit may have been the change from feeding human workers to allowing animals to feed themselves by grazing. The equation is altered if food has to be grown and/or carried to animals; extra feeding requires extra labour as well as organisation, so the equation outcome is contingent on the labour/cash/land cost of the fodder and the value of the extra work. Working-cattle users in arid regions often face the issue of poor-quality fodder and grazing at the end of the dry season, just at the time when animals are being set to the plough. Productivity can suffer severely as the animals lack the necessary strength, and once working, the cattle also lack sufficient time to graze or ruminate. Donkeys are generally considered to suffer less in productivity from these nutritional shortages; they commonly graze at night, unlike cattle, do not ruminate, and are physiologically more efficient processors of food.

The textual evidence in antiquity nevertheless underlines the importance of proper foddering arrangements for donkeys intensively used away from environments permitting foraging of any sort. This latter is found in donkey-caravan work throughout history and in urban equid management, as exemplified by McShane and Tarr's (2006, 2007) accounts of the city horse transport systems of 19th-century AD America, in which horses were considered as '"cyborgs", or as "a melding of the mechanical and biological"', and as machines for 'turning food into money' (McShane and Tarr 2006:365). While there are modern and ancient accounts of donkeys doing heavy transportation work being given extra feed, in general there is a marked divide in both the past and present between the low-key household use of donkeys common in rural Africa (relying on the donkey's ability to forage for itself) and heavy-duty 'commercial' exploitation of equid power.

Stall- or pen-feeding with fodder can be evidenced back to at least the early 4th millennium BC in Europe, and in Egypt and Mesopotamia to at least the 3rd millennium BC. Texts from late 3rd-millennium BC Girsu and Nippur indicate a ratio of 2:1 of barley sown to barley 'fed to the animal set to the seeder-plow' (Maekawa 1984:78), and similar ratios of wheat or legume seed to plough-animal fodder, though Sherratt (2006:346) reports a ratio of 1:1 in a text from Lagash (c.2400 BC). Stol (1995:195–6) reports wide variations in working-cattle barley rations cited in early 2nd-millennium BC texts, perhaps depending on the type of work being done, or the availability of grazing. In the 2nd-millennium BC Archives Royales de Mari, '[w]e learn from some letters that . . . they tended to feed the animals too much which resulted in disease' (Stol 1995:196). Texts from mid and late 3rd-millennium BC Mesopotamia (Mari, Beydar, Umma and Girsu – Heimpel 1994:11/14–21; Sallaberger 2014:341/347) set out the feed rations for donkey and hybrid teams pulling vehicles or ploughs, probably part of intensive operations in which grazing is not an option. Feed is recorded as being supplied to draught and plough donkeys and to donkeys and cattle in the winter; contextual information on the related burden of work and availability of grazing is generally lacking, and there is some debate among commentators about ration volumes, particularly as the donkey handler's wages may have been included in the ration allowance. The cost of fodder (mainly straw) is sometimes included in the cost of pack donkeys in later texts concerning the early 2nd-millennium BC donkey caravans between Aššur and Kaneš in Anatolia; Barjamovic (2011:37) makes the point that the inns on the route must have been large-scale growers (or buyers-in) of animal fodder.

Nutritional value of fodder

Another underrated element in the feeding model is the nutritional quality of fodder for working cattle and the implications of the need to add high-nutrition fodder crops to the agricultural mix. If food quality is poor, in hot regions oxen may not be able to maintain weight no matter how much they eat; and cattle given dry feed also need considerable extra water, at about four times the volume of feed (Renger 1990:273).

Professional working-animal handlers have long known that attention to nutritional value pays dividends. In 19th-century AD USA, owners of the many working horses in New York 'tracked fuel costs with great precision' (McShane and Tarr 2007:368) and discovered that feeding cornmeal rather than the same quantity of oats provided a lower work return: the reason was not known then, but it is now recognised that cornmeal lacks niacin. In highland Ethiopia there has been a trend towards 'plantings of sorghum over the more marketable tef because sorghum stalks and straw are more nutritious than tef straw' for their ploughing oxen (McCann 1984:5); archaeological detection of this might well form an indication of the presence of working cattle.

Meanwhile, animal herders in developing regions to this day have knowledge of which plants can correct mineral deficiencies and improve the animals' condition; Bangura (1988:297) reports for example that in Sierra Leone before ploughing, feeding of 'mineral supplements known as tupal [prepared] from a mixture of selected leaves, termite hill soil and common salt' to oxen obviates the need for extra fodder beyond grazing. Salt is another key nutritional element. In Burkina Faso, working animals are given sorghum straw or hay or compressed leavings from oilseed pressing, in addition to grazing; the farmers that I interviewed (JG Burkina Faso 2013) were unanimous that the outlay was worth it for the improved work-rate obtained. In 3rd-millennium BC Egypt fatted cattle were given extra nutrients by feeding green produce and dough (Brewer *et al.* 1994:86), and Stol (1995:196) notes early 2nd-millennium BC Mesopotamian texts listing beer-brewing residue fed to cattle for fattening. King (2014:13) reports from a study in Kenya that

> [i]t is common knowledge among herders that the quality of soils and the associated vegetation strongly control the health of grazing animals. This is hardly appreciated by palaeoanthropologists, who generally assume that all that is required is water and vegetation.

Sourcing of fodder

Crop residue is not very nutritious, but its use increases the return from a mainstream crop rather than turning land over to fodder-growing, which in highly populated areas is problematic. Extra labour is involved, though, in cutting crops low to obtain the straw and threshing and winnowing them to allow grain and straw to be used separately; varying heights of crop-cutting have been detected in prehistoric Europe, giving indications of straw use for fodder (Halstead 2014:51/176). Early threshing practices have attracted recent archaeological interest, with potential for detection of production of animal feed.

Extra nutritional feed can include foraged wild plants, specialist crops such as bitter vetch grown solely or mainly for animal use, and crops such as barley grown for both human and animal use. Early 2nd-millennium BC texts report cattle being fed grass, barley, beer-brewing residue, straw and reed (Stol 1995:195). Hay is less labour-intensive than specialist crops to harvest, though bulky to carry; it is more nutritious than straw but less than legumes, and it swiftly loses nutrition during storage (McCown *et al.* 1979:306). Wilkinson (1972:31) suggests that 'archaeologists tend to underestimate the amount of labour' needed for the fodder industry; the considerable organisation of harvesting and storage may well alter the balance of common models for working-animal use. Barrett *et al.* (1982:7/25) give a very relevant account of hoe farmers in Upper Volta (now Burkina Faso) adopting working animals and struggling in particular with learning how to conserve forage and formulate rations, and to reallocate their labour force to deal with forage and pasturing. Meanwhile, multi-use or specialist fodder crops may put pressure on the land available for human-use crops; although new land may be made useable with the new animal traction capability, the labour of sowing and reaping increases. Isager and Skydsgaard (1992:111) note too that in Classical Greece small-scale farmers were more likely to cultivate annual crops rather than perennials, which would require setting aside fields for fodder for several years.

Indications of the implications of the new foddering requirements accompanying the systematic use of working animals can be drawn from accounts of the massive hay industry that grew up in 19th-century AD New England

and Britain, with the rise of urban populations unable to grow their own fodder and making extensive use of working equids. Town horses' work was generally not seasonal, unlike hay supply, so a market system rapidly developed, with payment sometimes in manure brought back directly from the hay market. In America, from the time of the first colonies in the New England and Mid-Atlantic regions, '[c]olonists extensively farmed the natural saltwater marshes and freshwater meadows near the cities to provide salt hay (*Spartina patens*) for their livestock' (McShane and Tarr 2007:130), taking advantage of nearby water transport. However, upland hay (from alfalfa, timothy, clover and other crops) is much more nutritious, albeit much less convenient to transport. By the mid-19th century AD, upland hay had become New England's major crop and a key part of the state's economy; the state authorities urged citizens to use the salt hay locally for farm animals, bedding and manure mulching in order to free the valuable upland crops for export to the cities (McShane and Tarr 2007:130).

In developing regions today fodder demand in urban areas has an important effect on the rural economy, and fodder-growing and -carrying are huge industries. While overall the issue of transporting fodder for working animals may be self-solving in some circumstances, the location of fodder crops may well, as in the previous American example, be distant from the working animals consuming the feed. Isager and Skydsgaard (1992:105), referring to Classical Greece, comment on the complex organisation of fodder that must have been required for the oxen diverted to transport of building materials (likely to be distant from normal fodder sources) outside the ploughing season.

Grazing for working animals

Grazing issues

If grazing is to be used during the working season, the conundrum for the keeper of working animals is how to allow them to graze for long enough (and on sufficient nutritious material) every day to replace weight lost through work effort. A fundamental factor is therefore that the grazing area should be sufficiently near the working area; as touched on previously, arranging grazing for urban transport animals can be particularly problematic as they often work long hours. A further factor for cattle, especially, is the need to centre the grazing area on a water source, which restricts the grazing radius; in Mesopotamian regions today working cattle are generally grazed on cultivated land or 'riverine land with permanent pasture' (Oates and Oates 1976:120), with fodder a continuing issue. An Old Babylonian early 2nd-millennium BC letter comments, '"Perhaps the ú KIN . . . Attâ will tell you 'There is pastureland (*rītum*) of the oxen' – (well), the entire Euphrates is pasture land!"' (Stol 1995:184), though this was not open to all, and payments for using pasture land are recorded.

Kjaerby (1983:29) reports that in Tanzania 'plough cultivation has mainly developed in areas with agro-pastoral systems of land-use', with extensive grazing, though problems of course develop if cultivation increases beyond the carrying capacity of the grazing. Grazing may be controlled by potentially archaeologically-detectable practices such as tethering animals to moveable stakes, as still used in various regions today. Difficulties over concentrated grazing for working animals can be numerous; there are issues of crop protection and security of the animals (particularly valuable trained ones) from theft, and of overgrazing. Intensively grazed land also increases the likelihood of parasite infections. Farmers in arid regions such as central Burkina Faso and Nigeria today traditionally entrust their (male) cattle to nomadic herders for at least part of the year, when there is insufficient feed or labour for the feeding work, but herders' annual rounds take the cattle to distant areas at the critical working time; animals used for work therefore have to be managed intensively on-farm instead, with corresponding labour implications and the challenge of mastering the husbandry skills. Keeping working cattle distant from the farm and from their users also risks the animals losing docility (Reh 1982:108).

In northern Namibia, cattle are grazed on communal pastures as in many regions, but due to growing drought, the distances from pasture to farm are increasing, with the result that 'draft oxen may not be available for early land preparation' and the task is increasingly being given to donkeys as they can forage successfully on arid land near to the fields (Haufiku *et al.* 2004:176). Communal grazing near the working area can lessen the individual burden for owners but present its own management issues; Starkey and Apetofia (1986:15) report for example that in many cases of communally owned grazing in West Africa, 'the lack of agreement on grazing arrangements has inhibited the adoption of draft animals'. They add, though, that this may well be a cultural issue, as many villages in Nepal and Indonesia successfully cooperate in setting aside specific areas for grazing.

Grazing for ploughing donkeys generally offers fewer obstacles than for working cattle, but feeding long-distance pack donkeys presents particular issues, as working donkeys typically need to graze for at least six hours if supplementary feeding is not available (Förster *et al.* 2013:195). In the Old Assyrian texts dealing with the major pack caravans between Aššur and Kaneš in Anatolia in the early 2nd millennium BC, there are mentions of fodder (see previous examples) but few of the matter of grazing. Dercksen (2004:267) describes a reference to the unloaded donkeys being kept in a yard (as in Figure 5.5) or enclosed paddock (with the merchants paying the paddock owner), or at pasture outside the town, watched by a local herdsman.

Barker (2012) makes the illuminating point that in post-invasion Mesoamerica mule-caravan scheduling took account of time needed for the muleteers to plant fodder crops; from a different point of view in modern Bolivia, the timing and route of llama caravans is significantly affected by

Figure 5.5 Donkey park at a market in Takhar, Afghanistan

Source: © Stephen Blakeway 2012, by permission

feeding considerations, with some areas short of grazing or featuring poisonous plants, requiring the carrying of fodder (Nielsen 2001:166/170/185). Joubert (1995:130) reports similarly the considerations of grass quality in timing and routing of ox-drawn wagon-trains. Lattimore's (1928:49) description of arrangements for camel caravans in modern Mongolia also describes how caravan timing was arranged to coincide with the grazing season at the unloading point, where local landowners charged a fixed fee per camel monthly for grass and water; caravaneers concentrated on feeding up their strongest camels, leaving weaker animals to forage for themselves.

Physiological aspects of grazing

A key differentiator between the husbandry methods appropriate for cattle and for donkeys is their physiology in relation to feed and water. Both cattle and equids process their food by fermentation; in cattle this occurs in the rumen, and in equids in the caecum (a pouch-like element at the start of the large intestine) with an enlarged colon for storage. The rumen has limited capacity, and the rate of passage through the system is restricted by particle size, while the equid digestive system allows processing of bulky fibrous material: '[d]onkeys' tough digestive system can break down near inedible vegetation and extract moisture from food more efficiently' (Yilmaz 2012:17). Jones (2008:12) reports that while for (zebu) cattle almost half their intake must be high in carbohydrate and protein, for donkeys the figure drops to one-sixth, with much less need for total food and for high-nutrition food.

> Left to their own devices, donkeys consume dry grass, bark, leaves, twigs and roots of preferred species of plants, even creosote bushes in desert areas – not because they are hungry, but because they like them. . . . Donkeys can become ill on rich food such as alfalfa/lucerne and lush spring grass. (Yilmaz 2012:69)

Donkeys, with their strong jaws, are more ready than cattle to browse on woody species, and there are accounts of donkeys surviving on bark, fish bones, kitchen waste, paper and equid manure (e.g. Figure 5.6), though work performance improves with better nutrition. Cochin (1995:58) suggests that the donkey is 30% more valuable than the ox in terms of energy available for work in return for energy consumed in food, and Barrett *et al.* (1982:37) report from Burkina Faso that it costs four times more to feed oxen than donkeys.

Models of working-donkey husbandry systems in the present day are sometimes contradictory and driven by assumptions based on studies of other equids such as horses and zebras, by local prejudice or by mistaking of cause for effect: in South Africa, Starkey (1994b:4–5) reports from a survey in South Africa that

Figure 5.6 Donkeys foraging in rubbish, Mexico

Source: © Donkey Sanctuary 2008, by permission

> extension workers in several different areas stated that donkeys fed 24 hours a day, ate far more than an ox and were responsible for erosion and the degradation of the pasture. The frequency that such dubious statements were repeated suggests they may have come from some elementary text. . . . The donkeys were brought in because of the environmental problems: they were not the primary cause of them.

Pearson (1998:25) reports similarly that

> [d]onkeys are often accused of causing environmental degradation because of their apparent preference for short grasses. Account is not taken of the fact that donkeys are mostly found in fragile ecozones and in numbers exceeded by other stock.

The confusion may also arise from the ability of donkeys to alter their foraging strategy according to availability: donkeys – both domesticated and wild – are readier than cattle to select for greater nutrition when there is ample forage and/or they have already eaten well, while in times of hunger or poor grazing, they are uniquely well-equipped to process large amounts of low-quality forage (e.g. Smith and Pearson 2005:1).

There is ample support for the hypothesis that donkeys do and did not require anything like as much herding/pasturing as working cattle and that they therefore escape the 'grazing trap' described previously. In many African regions, donkeys are commonly released after work or in the dry season onto private or communal areas such as road-verges, or left to scavenge crop residue, dry grass and food waste. Donkeys are slower feeders than cattle and in the wild graze for longer, often at night. They are therefore often left to roam at night, though there are also accounts in several countries of their corralling to protect crops, to allow watering or feeding if grazing is poor, to allow rapid access during times of high demand for animal power, or to protect them from theft or predators. Cattle, too, may be confined at night if theft is a danger and herding labour is lacking. In these cases, the animals may need fodder, as I discuss previously.

As I hint at earlier on grazing issues, evidence of increasing donkey use over oxen can be an indicator of drought conditions. Smith and Pearson (2005:1–2) report for example from livestock census and weather data in Zimbabwe and Ethiopia that 'annual fluctuation of cattle and small-ruminant populations closely follow variations in annual rainfall, whereas donkey populations remain stable'; this is partly of course because excess cattle can be sold for meat. Unlike cattle, donkeys are native to semi-arid desert; again, unlike cattle, they can seek food over a wide area daily, and range at least 24 hours from a water source. Assessments of the length of time that cattle and donkeys can go without drinking are often speculative or anecdotal; both are said to survive at least 2–3 waterless days, but it is has been demonstrated by formal experiments that donkeys are significantly better able than cattle to withstand long periods between drinking without apparent stress or – importantly – loss of appetite, through a range of physiological and behavioural adaptations. Unlike cattle (and humans) donkeys can eat even dry food while severely dehydrated (Dill *et al.* 1980:975) and do not have to stand still to graze or to rest to ruminate after feeding; their energy cost of walking is also significantly lower than in cattle.

It should be borne in mind that African data on cattle physiology and behaviour often relates to animals with at least some zebu blood, who are more heat-adapted in various ways and likely to suffer less from water deprivation than pure *Bos taurus* animals. Taurines such as N'dama cattle have shown themselves to be fully adaptable to heat, but care is still necessary when drawing inferences about cattle performance in the Ancient Near East before the later advent of the zebu strain.

6 Transport

6.1 Donkeys for short-distance transport

Introduction

In 4th- and 3rd-millennium BC Mesopotamia and neighbouring regions, donkeys were adopted for local work in cultures commonly already featuring working cattle. The arrival of individual donkeys (always the basis, it should never be forgotten, of the more general phenomenon) was certainly related to pack use, fanning out from their north-eastern African regions of early adoption, and I would argue that their initial use outside long-distance pack was for local transport, an element largely neglected in Ancient Near Eastern archaeological commentary. There are surprising statements such as in Liverani (2006:18):

> in Lower Mesopotamia, as well as in the Nile valley and in alluvial valleys in general, the most economical and efficient means of transporting the harvest, from the fields to the threshing floors and from the threshing floors to the storage houses, was by boats. These could use the system of canals and river branches.

Algaze (2008:52–3) does recognise the limitations of water transport but only reluctantly accepts the likelihood of the effective operation of pack animals for this work. Local non-boat transportation, if mentioned, is too often associated with carts/wagons. Oxen employed for ploughing can, in theory, be used for on-farm pack work, but they have limited weight-carrying capacity per food intake and may require a fit adult handler; modern farmers may prefer to use donkeys, with or without small donkey carts, rather than employing cumbersome ox-wagons with their need for broad tracks.

Donkeys in modern developing regions typically perform the roles of wheelbarrows, bicycles or light motorcycles. Van Dijk (2011:¶1) provides a useful summary of how donkeys are typically used for short-distance transport:

- Providing primary source of income through direct use by transporting people and goods. Many landless people use their working animals to provide their main source of income because it is seen as a more reliable source of income than alternatives such as daily wage labour.

- Supporting agriculture activities such as transporting feed/water for livestock, cultivation activities such as ploughing and providing essential access to markets by transportation of agricultural products.
- Reducing the labour and drudgery of daily domestic household tasks such as collection of water and firewood especially for women and elderly. The use of working animals reduces the time spent on domestic activities and provides an opportunity for income-generation by women.

In Ancient Near Eastern models, the advent of ploughing and its social and economic implications are instead highlighted; yet in modern developing regions, ploughing donkeys in practice spend the majority of their time on farm or household transportation tasks as an essential part of their year-round return on investment. This suggests that the teams of donkeys described in a range of mid to late 3rd-millennium BC Mesopotamian texts as used for agricultural purposes are very likely to have been employed for short-distance transport as well as ploughing (Goulder 2018:83). Zarins (2014:189) points out that the 'plough' sign in the Uruk IVa Archaic Texts may sometimes denote farm work generally, rather than solely ploughing, just as now a tractor image often stands generally for farming.

Speedy transport of crops from the field to storage, processing or market has a marked economic effect as it hugely reduces harvest loss; heavy implements and manure can be transported to distant fields, and bulky fodder to farmyard animals. Raw materials (ore, clay) and construction materials can be conveyed further and in much greater quantities, altering the parameters for their sourcing and for manufacturing and construction sites. Short-distance transport has also been central to the operation of cities, from their earliest development. An agricultural surplus – considered a key concomitant factor to urbanisation – requires significant transportation facilities to local storage facilities and in from the hinterland, while the transport patterns of cities in present-day developing regions highlight the major daily inward traffic of firewood, construction materials and water.

Modern example demonstrates strongly how household use of donkeys today for the daily transport of fuel and water and of goods to or from market transforms the lives of women in particular, not only in relieving them of heavy burdens (with corresponding health improvements) but in providing free time for making goods to sell. The lowly status of donkeys, and their ease of use and low maintenance, permits their use by both women and the landless for income-earning, by taking goods to market to sell, carrying goods for others and hiring out their donkeys (Goulder 2018:83). I express in Chapter 6.3 my disappointment at the near absence of direct evidence from the Ancient Near East on use of donkeys for transport by women and other disenfranchised groups, which from the rich modern evidence may well have provided a social and economic transformative impetus. I look forward, though, to new insights from increased 'donkey-mindedness' in interpretation of relevant social and economic changes in the past.

Short-distance transport is defined for this book as centric, with the transporter (human and animal) returning to base, typically daily. The definition applies to the activity, not the animal, as a given donkey may be used for both long- and short-distance work: off-duty caravan donkeys might be used for local transport, and modern mobile pastoralists in regions of Africa use donkeys for both local forays, to exchange goods and acquire foodstuffs, fuel and water, as well as for long-distance trade and to carry household goods and equipment during moves. The camel-owning Tuareg, for instance, reserve their camels for long journeys and prefer to use donkeys (rounded up *ad hoc* from semi-wild groups) for short trips, as today a pick-up truck would be chosen instead of a large van.

Origins of short-distance donkey transport

The use of donkeys for short-distance transport outside their indigenous regions may have stemmed from the reuse of pack-caravan donkeys or those used by nomadic groups, perhaps those left in passing at settlements as today following injury, illness or giving birth. Donkeys might also have been brought by local traders and messengers returning from exchange points. Kreike (2010:95–8) reports how in north-central Namibia in the 20th century AD donkeys were initially brought in by returning migrant workers using them for transport of food and goods, and were subsequently imported in large numbers for their usefulness despite colonial disapproval; they then became increasingly adopted for ploughing as an alternative to cattle, which may suggest a pattern for the adoption of donkeys for farm work in Mesopotamia. Bulliet (1975:109) discusses how nomadic Arab camel breeders in historical times found short-distance transport work more lucrative than supplying camels for long-distance caravans; and the steady but flexible demand for transport animals for short-distance work – regular farm and household activities, *ad hoc* larger tasks – may also suit breeders more than the heavy but less frequent business of supplying long-distance caravans. From the user point of view, there are few entry barriers to short-distance donkey transport: donkeys can use human paths or tread their own route and can carry burdens directly on the back or in baskets slung with cords. Issues of feeding and work organisation are much less significant than with long-distance animals: typically, donkeys are left to browse locally and recruited as needed for work.

Short-distance transport today

Short-distance transport in villages and farms in the modern developing world is widely recognised as time-consuming and burdensome, in year-round terms far exceeding that of agricultural fieldwork (see the Waithanji

2014 survey findings that follow). Studies of transportation activity in villages in Tanzania and Ghana, employing the tonne-km measure (effort involved in moving one tonne one kilometre), demonstrated that three-quarters of the annual transportation effort occurred within the village; water, firewood and crops to the grinding-mill were the main activities, with water accounting for 70% of the tonnage, women taking the majority of the burden, and the processes taking up to four hours daily (Doran 1994:272–3). The time-consuming daily process of collecting fuel is perhaps particularly underestimated by archaeologists more familiar with temperate, forested regions.

Transporting goods for others by donkey is commonly reported in studies of developing regions today as an important source of income, particularly for women and other landless and disenfranchised individuals, but also for farmers aiming to maximise working-animal utilisation in slack seasons, with donkeys seen as productive assets in the same category as land and other holdings. Transporting fuel and water for financial gain is a staple in modern Africa and elsewhere; this trade is likely to have ancient antecedents, as instanced by the common association of donkeys with water-carriers and woodcutters in the texts from 2nd-millennium BC Deir el-Medina in Egypt (Janssen 2005; Janssen *et al.* 2003:26). Payment for transport services is often in kind, as in antiquity, including repayment of obligations or storing up of goodwill.

Women and the impact of donkey adoption

Transforming women's lives

In societies where cattle ownership and usage are prestige-related and largely the preserve of males, women often have considerably more cultural access to lowly donkeys. Throughout history and to the present day, donkeys are commonly despised – notably by non-users – and regarded as low-status: an FAO working paper on animal traction worldwide notes that 'donkeys have fewer associations with masculine power than most other work animals' (Starkey 2011:27). A theme in numerous published modern agricultural and socioeconomic overviews and workshop papers is of such gender-neutral attribution allowing women to employ donkeys without the cultural constraints attached to cattle.

The carrying of burdens is known as the traditional role of women in many cultures (Figure 6.1), and NGOs and agro-economic organisations, in African regions in particular, have focused significantly in recent years on the overwhelming benefits of donkey use for easing the work of women. In a major quantitative NGO survey in Ethiopia, this easing is found to be the single most important social contribution of donkey ownership (Admassu and Shiferaw 2011:8). As well as reduction of hardship, the delegation to donkeys of heavy load-carrying and manual cultivation work reduces reproductive and other medical problems in women; among mobile groups, donkeys can transport small children, lessening constraints on family size

Figure 6.1 Bethlehem woman carrying water in a goatskin (c.1934)

Source: www.loc.gov/item/mpc2010004464/PP/. Library of Congress, Prints & Photographs Division, G. Eric and Edith Matson Photograph Collection. Photograph in the public domain

(Mitchell 2018:36). Donkeys correctly employed are docile and easy to handle by women, who can also pass on to children daily tasks such as taking a donkey to collect water. Occasional barriers are reported, as in some West African societies, where women have access to donkeys but still carry fuel or water themselves (e.g. Starkey 2011:26–7); the underlying reason may in fact be economic, as reported by Doran (1994:275): '[h]ouseholds had to choose between conserving the energy of their animals or that of their women; in many cases the choice favoured the animals'.

The main work typically transferred to donkeys is household and farm transportation, but their use for cultivation can also provide new independence for women constrained from working with cattle. Donkey ownership also allows women (and other subordinate categories such as landless males) to conduct income-earning activity such as taking produce and home-manufactured goods to market (e.g. JG Burkina Faso 2013) and carrying goods for others. Ewers (1955:308–15) gives a parallel among Native Americans: with the advent of the horse, women in some cultural groups – the chief load-carriers – enjoyed higher social status and were freed to engage in other tasks such as craft production. Marshall and Starkey 1998:33, in a paper at an international animal-traction workshop in Ethiopia, report that in modern Africa

> [r]ural women are often extremely busy with household work ('reproductive tasks'), including the transport of fuel wood, water, food and children. . . . Donkeys may also offer women the opportunity to enter the usually male-dominated income-generating ('productive') sphere. Donkeys can be used to transport materials for financial gain (e.g. transport of water, wood, forage, produce for market). This helps women (as well as men) gain money, status and, as a result, power. . . . Male-dominated societies often consider women as low-status citizens, while donkeys are low-status animals.

Meanwhile, the loaning of donkeys to neighbours and relatives is a valuable route, for women and more broadly, to respect and status in the community.

A striking feature of modern use of donkeys in daily village and farm life in developing regions is the divergence between usage by men and women, with much direct commercial activity still often lying in the hands of males (e.g. Pearson *et al.* 2001:25). Accounts from present-day Iraq, Kenya and Ethiopia indicate that men are notably more likely than women to be using donkeys for commercial transport for income, especially in urban areas, with women more likely to use them for rural household transportation. However, Hart (2011:18) argues cogently that the subordinate position of women in sub-Saharan Africa made them 'quicker to exploit the commercial freedoms of the neoliberal international economy' and more able to operate under established (male) polities. When Boserup (1990:14) argued that '[w]omen's work, women's fertility, and women's role in the family and in society at large are radically changed by economic development' – notably that involving new technology – she is referring to the negative effect, resulting in women's exclusion; the advent of the donkey in modern developing regions is therefore a guerrilla technology, reversing the effect and undermining established cultural rankings by proving the most useful to the less-privileged of a society.

Invisible women in Mesopotamia

Can conclusions be drawn as to the possible consequences for women in Mesopotamian society of donkey use, notably reduction of time spent carrying household and farm burdens? Commentators modelling social impacts of the adoption of working animals have commonly employed Marxist-related constructs for discussion of gender implications, such as property ownership, inheritance, labour use and power, rather than direct daily impact. These models, too, with their influence on Sherratt's original secondary-products hypothesis, focus strongly on the adoption of ox-ploughing, leaving unexamined the role of donkeys.

The position of Mesopotamian women in the 4th-3rd millennia BC can be glimpsed only sketchily. Elite women are occasionally referred to in 3rd-millennium BC texts as holding influence and power (Asher-Greve 2013:360, in her useful synthesis of evidence about women in Mesopotamia); but texts and depictions of low-class workers from the 4th millennium BC on are commonly non-gender-specific or ambiguous, though there are nevertheless strong indications of low-class women enduring poor working conditions, receiving smaller rations than men and not benefiting from land allotments (e.g. Adams 2008:11; Asher-Greve 2013:368). Adams (2010) reports on references in the late 3rd-millennium BC Garshana Archives to non slave female brick-carriers, possibly commuting in daily from surrounding settlements; bricks are of course commonly conveyed today by donkey (Figure 6.4). Asher-Greve (2013:359) argues from anthropological research that women's lives in farming communities differ fundamentally from those in cities, placing limits on the value of urban textual evidence in Mesopotamia with its largely rural population. Adams (2008:11) concludes from examination of the textual statistics of women workers that 'a very high proportion of women were primarily engaged in their own subsistence pursuits and resided with their own families'. This suggests new insights – only speculative with the present evidence from antiquity – on the social implications of the advent of systematic donkey use on the Mesopotamian 'private sector'.

Short-distance transport in urban contexts

The relationship between agriculture and urbanisation, notably in 4th-millennium BC Mesopotamia, is commonly discussed in terms of plough agriculture as an enabler of crop surpluses, whether as a cause of population growth or an effect. Some separate references occur to the importance of long distance transport for the growth of complex societies and cities (e.g. Zarins 2014:198), but very few focus on the contribution of short-distance transportation to the day-to-day logistics of supporting an urban population. Zarins (2014:188), in his important work on equids in the 3rd millennium BC, emphasises appropriately that '[f]rom almost the beginning of the cuneiform record, when the equid presence can be recognized, these animals contributed significantly to the Mesopotamian economy', but makes no mention of the central role of donkey short-distance transport on farms and in villages. He assumes that equids 'were used primarily as draft animals in a myriad of activities associated with the city as well as the country', with separately the development of long-distance pack caravans. He notes that the economic texts 'tend to emphasize a routine pattern that involved farm work and remained relatively silent on such functions as warfare and riding'. Donkey farm work, as I note throughout, consists principally in history and to the present day of short-distance pack transport work, with perhaps seasonal ploughing; other traction, and riding, were very likely to have been minority activities for a now large body of working donkeys involved primarily in short-distance pack transport, their main role from then until the present day.

Transport from field to farm and to consumption location has always been available in human form, but archaeological models of surplus-provision rarely address the practicalities of dealing with the major increase in bulk for transfer from the hinterland to newly urbanised nodes. Branting (2010:51–7) describes detailed models for 'pedestrian transportation simulations' (p. 53) applied to 1st-millennium BC Kerkenes Dağ, for analysis of urban movement patterns such as bulk grain import; the models go down to the level of human stride length and pedestrian age, sex and footwear, but remarkably the operation of a similar system for animal transport is only mentioned in passing. Reference to short-distance transport into urban areas is generally to agricultural products, whereas overwhelming evidence shows the key daily items conveyed year-round into urban areas in modern developing countries to be fuel, water and construction materials.

In a census of donkey and human transportation in and out of Kano in Nigeria (Robinson 1977:13), donkeys' main burdens were firewood inwards and dung out to the fields. Accounts of 19th-20th century AD urban equids in the United States and Britain underline the problem of dung removal (e.g. McShane and Tarr 2007:ix); it may be collected by farmers in return for fodder supplies, which are in themselves a major issue for urban working animals. There is no direct Western-world parallel as the waste from fuel entering in vehicle fuel-tanks is emitted as exhaust gases; in its place is the major outward flow of household and industrial 'rubbish' for landfill, a concept hardly known in antiquity.

Reports such as Admassu and Shiferaw's (2011:27) working-equid survey results in Ethiopia demonstrate that in modern developing regions, use of donkeys in urban and peri-urban areas is commonly intensive and potentially lucrative, for transport of grain, firewood, water, construction materials and fodder; this is a useful occupation for landless individuals. In modern Ethiopia '[a] large part of the people and of the economy of Addis Ababa depends on donkey transport for the movement of grain from wholesale centres to retail outlets and households' (Zenebe and Fekade 2004:69), with several thousand donkeys employed daily at the vast Yehil Berenda grain market (Figure 6.2). As I note before, there is evidence in some regions of the majority of urban commercial donkey transport being conducted by men,

Figure 6.2 Loading 100 kg grain-bags in Yehil Berenda grain market, Addis Ababa
Source: © JG Ethiopia 2014

though in modern Ethiopia, women and children have some hold in the business of urban water-selling (Pearson *et al.* 2001:26).

In Chapter 2.1 I cite Brodie (2008:304) as arguing that new modes of short-distance transportation to supply large centralised demand might have been a necessary context in which urbanisation flourished, albeit not an initiator. Cities then as now became heavily dependent on their supply systems, with predictability of supply of foodstuffs and raw materials even more crucial than quantity. This could shed light on the 'accounting fictions' (Chapter 6.3) detected in later 3rd-millennium BC texts on agricultural output: urban authorities demanded firm predictions of supply, and – looking ahead to the eventual implosion of these urban systems – their received information may have diverged increasingly from the actual deliveries (by donkey).

Analogies between urbanisation in 4th-millennium BC Mesopotamia and modern situations are necessarily tentative and must rely on the resilience of basic forces; but insufficient attention has been paid to the practical workings of modern major cities. Tarr (1999:434) encapsulates the inescapable need of cities for local essential supply systems:

> [o]ne way to think about cities is to conceptualize them as energy systems – as entities that require flows of energy for a wide range of purposes including heat, light, and power. Over time, the sources of these

energy flows have changed from human to animal power, and then to steam, electricity, and gas.

Marginal land

The nucleated settlements and large centralised farms of 3rd-millennium BC Mesopotamia presumably raised the issue of a farm-to-field transportation system, perhaps resolved through animal transportation. Farmworkers may walk several kilometres daily to distant fields, but there are limitations and diminishing returns. While walking such distances daily is nevertheless common in many developing regions, for example if there are security dangers in living outside settlements, it implies loss of working time at least. Isaakidou (2008:104) suggests that adoption of draught cattle at Knossos was generated by a need to carry manure to more distant fields as the crop-using population grew; Brodie (2008:302–3) argues that the Bronze Age population growth in the main settlement on the Greek island of Melos could only have been achieved by 'an agricultural system that relied on animals for transport', allowing the use of distant fields, and that the transport animals could reasonably be assumed to be donkeys. He continues (Brodie 2008:304)

> once it had become viable to farm larger areas of land at locations further removed from settlements, residential options multiplied . . . [allowing] flexibility of choice in both settlement location and degree of household aggregation.

As well as social decisions, economic factors were affected, as farmers would then be freed to choose field locations for reasons other than proximity: Hassig (1985:264) records about Conquistador-period Mexico that with the advent of animal transport '[a]reas of agricultural exploitation developed with more concern for favorable production than for transportation ease'.

The issue of location of cultivable fields, in the face of population growth or drought, is regularly aired in modern Africa and elsewhere. In western Uganda, for example,

> [i]ncreasing human population and the resultant pressure on land has forced many farm families to cultivate in distant places far removed from homesteads where there is relatively abundant free land . . . [resulting in] failure to quickly transport harvested crops to places of storage and or markets. Consequently, high post-harvest crop losses were experienced through vermin, rain, mouldiness or aflatoxins, theft, termite damage and spillage. (Ayo-Odongo *et al.* 2000:211)

This problem was solved by the introduction of the donkey for speedy transport of crops to storage and transport. As I note in Chapter 5.2, in northern

Namibia drought conditions have resulted in greater distances between viable pasture for grazing cattle and available land for cultivation, leading to an increase in use of donkeys for ploughing (Haufiku *et al.* 2004:176).

Items carried short-distance

Items carried locally naturally vary according to conditions: in dry zones, water and fuel is transported for longer distances, while more rainfall increases the volume of agricultural produce to be carried; the use of animals for ploughing in itself increases the agricultural load to be conveyed. The carrying of firewood and water, though, is the overwhelming mainstay of short-distance donkey use in sub-Saharan Africa, mentioned more often in modern studies than any other task. The use of donkeys for transport of crops from the field for threshing, storage or selling is also ubiquitous; the speedier transport to storage or market reduces the risk of deterioration and theft as I note previously, so farmers without transport may find it worth their while to pay for carriage. However, transportation of crops is of course seasonal.

Donkey transportation of crop residue and specialist fodder is common in modern times (Figure 6.3); crop residue such as chaff is also used in ceramics and construction materials.

Figure 6.3 Donkeys transporting straw in Addis Ababa, Ethiopia

Source: © Noah Snyder-Mackler 2010, by permission

Dung – a heavy load to carry – is a common burden for donkeys, from fields to farms or *vice versa*, for fertilising or for fuel or building. Manure is also exported from urban areas, as I note previously. The modern equivalent, fertiliser, is also commonly transported by donkeys from distribution centres and to the fields; Starkey (2011:11) describes how '[w]ork animals create synergy in nutrient cycles, farming and marketing systems', helping to circulate manure and the resulting harvest.

Donkeys are extensively used in modern Africa, by sedentary and mobile groups, for the transport of items for sale at markets (farm produce, products gathered from the countryside, and manufactured items such as pottery and beer), and for carrying purchases home from market. Liverani (2006:44) argues for the existence of local markets in Mesopotamia below the purview of the recording authorities, for household-level exchange of agricultural produce and for example pottery. The use of donkeys allows access to more distant markets and transportation of heavy and bulky goods; this results in new goods exchange nodes, as at the Ethiopian Yehil Berenda grain market which I mention earlier; this escalates into another process, unsung in archaeology until recently but now increasingly acknowledged, of the capillary feeder-supply of goods to arterial routes and 'railheads' – ports, collection points for long-distance caravans.

Waithanji (2014:2) reports from a detailed survey in Kenya that the greatest contribution of working equids is

> hauling produce through the "first mile" of the logistical journey of the value chain. The first mile is constituted by the distance between the point of production and the point of efficient transportation using vehicles travelling on all weather roads to the consumer markets. In parts of Kenya, the first mile has been established to be about 1.5 to 13 km, which constitutes 0.4% to 10.6% of the entire logistical chain . . . The study also revealed that 20–37% of all transport costs are expended in the first mile. . . . Women, more than men, tend to be concentrated at this first mile, which constitutes part of the bottom of the value chain where labor intensity is high and capital demands low. At this point too, 30–60% and sometimes, even 100% of post harvest losses due to transport challenges are documented.

As I suggest in Chapter 2.1, pastoralists may well have been early adopters of donkeys in Africa for carrying dwellings (especially the vital poles), furnishings and household possessions during their seasonal moves (Figure 6.7). Among sedentary groups, a change of construction material (or of source) might give an indication of animal transportation, as in 4th-millennium BC Europe where an increase in use of heavier timbers and other house-building materials became apparent (Pétrequin *et al.* 2006:11). Modern and historical accounts of donkeys in Africa and Asia commonly report the transportation of construction materials (mud, dung, grass, sand, stones, poles, straw, bricks) for building or repairing housing and other structures (Figure 6.4), sometimes for income; soil also needs to be moved from excavations for irrigation, etc. In arid regions timber in particular, aside from use for fuel, is carefully collected, curated and re-used for building, boats, equipment such as looms, and agricultural implements and tools; donkeys also commonly carry the implements themselves to and from fields and workplaces.

Only limited attention has generally been paid to modern and ancient transfer of heavy or bulky raw materials for manufacturing processes, such as clay, ore, charcoal, wool and flax. Changes in manufacturing site locations, for example further from sources of raw materials and more convenient for onward transport, might indicate an improved transport technology, as with the development and organisation of copper-working in the Negev in the southern Levant from the late 4th millennium BC. Refuse also comes into the 'materials' category: the concept of 'rubbish' is a modern Western one, and in developing countries even items not recycled in the household are sorted and reallocated, particularly in urban areas, by specialist collectors such as the Zabbalin in Cairo and marginal groups in Amman, and commonly transported by donkeys for use or sale.

While this study excludes discussion of riding, the transportation of people and animals (young, infirm or dead) also qualifies as short-distance transport. Donkeys are used by pastoralists in their seasonal moves to transport the young, the old and the disabled, and in both sedentary and mobile communities to carry the sick or injured to where they can be treated, as well as bodies for burial. The adoption of remote burial centres might give

Figure 6.4 Carrying bricks in India
Source: © Stephen Blakeway 2014, by permission

indications of the adoption of animal transport, as in the late 4th millennium BC in the northern Negev (which I note previously as an early location of donkey use) where burials took place in caves and at hilltop cemeteries. Injured or young animals are also transported by donkey, plus small stock such as chickens, and carcasses are carried by working animals from the hunt or slaughter.

6.2 Other uses of working animals: an overview

Carts

Carts, wagons and sledges can be pulled by humans, but their systematic use is generally associated with traction animals. Köpp (2013:111) reports on an Egyptian 12th-Dynasty (2nd-millennium BC) depiction of a huge statue pulled on a sledge by humans, suggesting that '[a] possible reason for the use of men rather than oxen is that people could be better directed and show quicker obedience to orders than oxen'. The development and diffusion of wheeled vehicles in the late 4th millennium BC in the Near East and Europe have formed the subject of vigorous pre- and post-Sherratt debate, with Littauer and Crouwel's work central to the discussion; I do not aim to contribute to this, particularly as there are technological obstacles to comparison with modern examples. I provide a summary for general context.

Sherratt based his original traction 'complex' model (1981:263) on a linkage of plough and cart technology, taking an analogy from use of oxcarts for household moves in the Central Asian steppes. He subsequently discarded this parallel (Sherratt 2003:243), concluding that early wagons may have been represented as part of a power and status display on monuments and in burials, but in practice were little-used for everyday load-carrying. Claims for models and traces of wheels from the 4th and early 3rd millennia BC in Mesopotamia and Europe are sometimes debatable, and Sherratt (2006:333) takes a robust view on excluding precisely-dated artefacts of dubious purpose and poorly-dated sites of clear purpose. The iconic Mesopotamian Eanna IVa pictographs (c.3200–3100BC) show sledges and possible wagons, sledge-like but with two discs on the side view, perhaps indicating wheels or captive rollers.

Mesopotamian 3rd-millennium BC evidence of wheeled vehicles is patchy (e.g. Littauer and Crouwel 1979:13), and wheeled transport in antiquity was indeed not suited to competing on long-distance work with river-boats or pack caravans. Dercksen (1996:64–6) records a few references to wagons in the Kaneš texts, but seemingly for onward transport of goods on arrival in Anatolia by pack-donkey caravan. The recommendation for a wagon for farm work in the 2nd-millennium BC *Farmer's Instructions* (Black *et al.* 1998–2006d:line 92) may relate to an ideal rather than to common practice in Sumer. Analyses of early representations suggest that use of donkeys for traction may have been initially constrained by unsuitable use of harness perhaps designed for cattle; plentiful examples of counter-productive harnessing methods for donkeys persist today.

Littauer and Crouwel (1979:27), supported by Zarins (2014:193), report that no Sumerian text link is evident between equids and wheeled vehicles, and representations are generally rare until the mid 3rd millennium BC; the exact type of equid indicated is the subject of debate, though it is generally recognised that depictions of the mid 3rd millennium BC are donkeys or onager-donkey hybrids. Inevitably, little can be deduced from elite representations as to the role of cattle or donkeys for traction outside ceremonial occasions. They could be an irrelevance, as if our present culture were to be judged by illustrations of gilded carriages at state occasions, but as Sherratt (1997:33) points out, 'it is important not to fall into the trap of assuming that 'prestige' equals 'useless for practical purposes''. Littauer and Crouwel (1979:31) suggest that equids became more widely used for traction but that traction cattle were still depicted in art and found in elite graves as the traditional though not necessarily day-to-day norm, just as plumed horses may be used today for a hearse.

Sledges and travois

Threshing

Threshing by sledge is often more rapid and efficient than manual methods for producing grain and particularly chaff (chopped straw), though other threshing technologies persist to this day. Chaff is laborious to cut up manually and is of key value for ceramic and mudbrick temper, animal feed, bedding and fuel; its advent is visible archaeologically, for example in dung residue. It is possible that cattle were first used systematically for work in the threshing process; there is growing evidence in Mesopotamia from the 8th millennium BC onwards of sledge-based threshing, perhaps reflecting the new presence of free-threshing cereals or the need to feed animals. Humans can and do pull threshing-sledges (Anderson 2006:310), but the extra power and weight of animals – notably cattle – and their sharp hooves make a critical difference.

Donkeys and cattle are used for threshing to the present day, for trampling or pulling sledges. There is little textual mention of donkeys threshing in Mesopotamia until the late 3rd millennium BC, though this by no means precludes their common usage in practice: in the 3rd millennium BC in Egypt, donkeys were common threshing animals. By the 4th millennium BC in Mesopotamia, threshing-style sledges appear to have attracted some ceremonial significance, as evidenced by seal impressions and the funerary sledges in the Ur and Susa graves of this period. In modern times, Halstead (2014:173–7) describes ceremonies associated with threshing in modern

Europe, but highlights the archaeological evidence that workaday threshing was occurring in the fields contemporaneously with the ceremonious threshing-sledge use depicted.

Sledges and travois for transport

Sledges and (less often) travois are commonly used today in parts of southern and eastern Africa and elsewhere, though they are officially banned in some areas as they are considered to cause erosion (e.g. Starkey 1994a:74). There is some correlation with use of oxen for ploughing – unsurprisingly, given the force needed to pull sledges – and with scarcity of donkey carts.

Travois are a simple construction of crossed poles or branches fastened at the animal's neck and dragged by flexible traces with a load. Unlike sledges, there are no illustrations of or textual references to these in the Ancient Near East, but this reflects their lowly, rural function. Experiments with newly adopted working animals may have led to their development: bushes are dragged by animals in modern Africa in a way similar to travois use. The cost of a travois is minimal for owners of working animals: a survey in a region of Tanzania showed that most ox-owners own a travois-like construction of tree-branches (Kilemwa 1999:70).

Sledges and travois are useful and sometimes essential for a range of items, including transporting the ill and deceased. They are said to be especially valuable in sandy/dusty regions and on steep slopes (e.g. Halstead and Isaakidou 2011:62); travois are harder to drag on firm surfaces, but they are useful on rough ground and perform almost as well as runner-sledges on soft ground (Dennis and Smith 1995:120). Direct dragging, of timber in particular, is also used today and was employed in antiquity, for example from the Lebanon, Taurus and Amanus mountains to the Mediterranean and Euphrates for onward travel by water.

Stationary power

Working animals have also commonly been employed for providing stationary power, in Classical times, the recent past in the Western world, and developing regions today (Mitchell 2018:123). This can take the form of wheel power (raising water, milling, pottery-throwing, spinning, butter-churning, dough-kneading), hoisting (wells and irrigation, construction, loading of boats) and pressing (hay, cotton, clay, oilseed, olives, grapes). Bulliet (1975:217) notes that when wheeled transport fell out of use in Egypt in the early 1st millennium AD, non-transport wheel use continued for stationary power; as Mitchell (2018:166) points out, Bulliet's hypothesis of the post-wheel predominance of the camel takes almost no account of the donkey or mule, common sources of stationary power to this day. In pre-conquest Mesoamerica wheels were also commonly used for non-transport functions such as calendars and toys, and in Mesopotamia the pottery slow wheel was known from the Ubaid. As with carts, ancient/modern comparison presents technological issues and I do not address the subject.

Oxen, camels and other animals for short-distance pack

My chapter on short-distance transport focuses on animals used systematically for pack work in Mesopotamia and the southern Levant in the 4th and 3rd millennia BC, and so on donkeys; but for completeness I add a brief note on other animals used at times throughout history for pack, both long- and short-distance. Certainly experimental use of other animals has been likely since domestication, as indicated by the well-known sheep and bovid figurines from the southern Levant in the 4th millennium BC. Some domesticated species have been locally useful where other options are costly or unavailable, and there are instances of *ad hoc* use of tamed animals from the wild, which was indeed very probably the route of adoption of the main pack animals.

I have excluded from my study the use of elephants and dogs for pack and traction, as their size and circumstances of use have only limited relation to cattle and donkeys; I also exclude horses and mules.

Cattle for pack

Cattle, while valuable for traction, can carry much less per liveweight than donkeys, particularly in difficult terrain: they have poor stability and balance, they are slow, and unlike the desert-adapted donkey need significant care in terms of feed, water and rest. Egyptian prehistoric rock-art includes representations of cattle with packs tied to their horns, though it was likely that the work in due course shifted to donkeys, which were better-suited to travelling the growing distances between habitable places as Saharan aridity grew.

Lindblom (1931) gives examples of use of oxen for pack in Africa, in pack-trains or by nomadic groups using oxen for transporting their tents and household equipment; while in India until the era of mechanisation, the Bangara caste were traditionally long-distance transporters of grain and salt using large caravans of pack-bullocks. More recent accounts report on some continuing usage of pack oxen, mainly in Africa.

Camels

There is much speculation about possible domestication of the one-humped (dromedary) camel for pack earlier than the 2nd millennium BC, possibly in Arabia or Iran, though perhaps initially for meat and milk; meanwhile the colder-climate two-humped Bactrian camel was probably domesticated in the 3rd millennium BC in north-eastern Iran. Wild camels may have inhabited the Levant and even Mesopotamia in the 3rd millennium BC or before, but alleged 3rd-millennium BC depictions in Mesopotamia and the Levant are likely to have been drawn from hearsay.

The camel can carry up to 300 kg depending on the distance; accounts of daily caravan distances vary but converge around 40 km. The one-humped camel has clear advantages over donkeys in sandy desert regions with few watering-points, though they are unsuited to rocky, steep terrain; their roles today remain complementary and even interchangeable. The keeping of camel herds, meanwhile, is different in style and purpose, with prestige, milk and hair playing a part apparently largely absent in donkey-keeping. As with donkeys, camel reproduction is comparatively slow due to late puberty, seasonal oestrus and long gestation.

Goats, sheep and regional working species

Goats and sheep have been used for pack in the lower Himalayas and for pulling carts in Latin America, Australia and Europe as well as for a range of other specialist *ad hoc* traction tasks; large goats are used nowadays for pack-carrying for hikers in America. Cart-pulling goats are evidenced in Late Minoan Crete, and Ovadia (1992:24) notes a mediaeval Arabic term for a leader ram carrying the shepherd's provision bag. Certainly it is plausible that early *ad hoc* experiments in the southern Levant hills, as indirectly hinted at by Chalcolithic ram figurines bearing items on their back (Epstein 1985:53 among others), opened the concept and aided acceptance if not domestication of the donkey in the region.

Among animals elsewhere of a similar pack capacity to donkeys and cattle, reindeer have been used to the present day for pack and transport in generally low-management systems in Lapland and Siberia. There are also instances of elk being domesticated for traction in the former Soviet Union. Yaks are used to the present day for transport in the Himalayas, as are the yak-cow cross the dzo or zopkio, bred for supply to caravans by specialist transport-animal groups. In Mesoamerica the Incas operated llama trains between the highlands to the coast, and llama caravans continue to this day in Bolivia (Nielsen 2001); llamas carry 45 kg at most and cover less ground daily than porters carrying similar loads, so although they offered the economic benefits of meat and wool, they were not as technologically advantageous as the donkey in the Old World. In Africa, zebra (Joubert 1995:135) and eland have been experimentally used for traction (including breeding zebra-donkey crosses (Clarence-Smith 2016:119)), but this has never become popularly or commercially successful.

6.3 Long-distance pack-donkey caravans

Introduction

In contrast to other sections of this book, where plentiful modern material informs interpretation of the fragmentary findings in antiquity, this chapter does not provide detailed analysis of its subject. The reason – the invisibility of pack-donkey caravans in antiquity and to the present day – is a finding in itself and is a sharp reminder of the likely extent of activity unrecorded in texts from the late 4th millennium BC onwards. There has been strong interest throughout history in conducting profitable activities under the radar of authorities; donkeys are ideally suited for moving of goods outside official inspection, via long desert or mountain detours and back routes as to the present day. From the thin evidence available, modern donkey caravans in Africa actively avoid passing through most settlements, for speed and probably fearing taxation or other authority intervention, homing in instead on large (sometimes 'pop-up') regional markets. Studies of more recent pack-caravan activities also underline the importance of protection against banditry, which overlaps considerably with local tolls; security can be achieved through guards and alliances but also through simple avoidance by routeing through difficult terrain.

Modern caravan drivers worldwide seek invisibility not only from tax-hungry authorities and preying bandits, but also from the unwanted attentions of NGOs who would regulate and publicise their treatment of working animals. Other barriers to information on pack caravans are encountered in Latin America where recording of working-animal activity is often funnelled through veterinary services, who are unlikely to be visited by caravaneers. Few researchers from any discipline have the time and resources to follow modern caravans, with valuable exceptions found in Helina Woldekiros (pers. comm. 2016) who accompanied donkey and camel caravans in NE Ethiopia as research for her PhD, and Nielsen (2001) who accompanied a llama caravan in Bolivia, reporting in detail on the caravaneers' motivations, activities, routines and the archaeological implications.

Laden donkeys can negotiate rocky, precipitous paths, and advantage was taken of this by Assyrian merchants in the early 2nd millennium BC (see my later section on the Kaneš texts) for taking difficult back-routes into Kaneš to avoid tax-payment on their goods. Modern entrepreneurs in Africa and elsewhere, including from Turkey into Iran and Iraq and between Lebanon and Syria, have taken this a step further, making profitable use of donkeys' excellent memory for routes, good night vision and ability to work unsupervised by sending large trains of donkeys (and mules) unaccompanied across national borders laden with smuggled goods, through desert, mountains or dense vegetation where capture is unlikely.

The invisible long-distance pack donkey today

Most descriptions of caravans are therefore garnered from accounts of travellers in recent centuries, as with the examples to follow, and are highly anecdotal; considerable original research would be required for an authoritative account. Riemer and Förster (2013:36–52), as part of their valuable work on modern caravans *Desert Road Archaeology in Ancient Egypt and Beyond* (Riemer and Förster 2013), list some of the few studies extant on northern

African and Latin American donkey caravans, many focusing largely on geographical information; some caravans also continue in mountainous areas of China and Nepal. The Riemer and Förster team themselves in their local investigations in northern Africa only encountered one modern pack-donkey caravan, by accident (Förster *et al.* 2013:211), for informal interview. They underline from notes of this encounter the many variables in any statistical analysis of time and load: condition of the donkeys, terrain, climate, duration, daily travel routine, experience of the animals and drivers, and particularly the available feed and water along the route, which dictates the hours travelled.

An account by Cable and French (1950:159–60) vividly encapsulates the nature of medium-distance donkey pack caravans in western China in the 1920s/30s AD.

> One man, or at the most two, will drive twenty donkeys, riding behind them, shouting incessantly, and never letting them slacken to normal walking pace. . . . The donkeys are small and cheap, so he is careless of life and sacrifices them in large numbers to his passion for speed and his reckless output of strength. He will use dangerous short-cuts over which no other class of transport-man will venture, and in bad weather many beasts die by the roadside.

For longer-distance journeys, the few accounts of donkey-caravan operation in Africa offer glimpses of the complexity and organisation involved. The long-established salt caravans in north-eastern Ethiopia (Figure 6.5), dating from the 6th century AD and described in Chapter 4.1, fed important salt-bar exchange currency into markets, for purchase of goods and livestock also brought long-distance.

Figure 6.5 Salt caravan from the Danakil Depression, Ethiopia

Source: © http://theonearmedcrab.com, by permission

Vast caravans were assembled, by small-scale traders allying themselves with the large merchants: '[w]hile the going was good the merchants tended to divide up to avoid too much concentration in one area, while in periods of insecurity they stuck together to defend themselves' (Pankhurst 1968:352), and '[h]alting places were often in isolated areas to avoid disputes with local populations' (Pankhurst 1968:347). There were no caravanserais, but temporary stone walls were built for protection and storing of packs; animals were led away from the camp to graze, then tied to stakes, and sometimes protected from predators by a thorn fence.

In a similarly complex arrangement, since several hundred years ago and continuing today, a figure-of-eight donkey-caravan route has operated whereby the Yarse of central West Africa convey cotton goods and cereals northwards to Saharan markets such as Timbuktu and return with salt slabs; some of these are then taken by different (and fewer) caravans to southern West Africa; there in Ghana the donkeys are in high demand among Hausa traders, who use them to carry kola nuts (highly valued for chewing) back to the Saharan region (Binger 1892; Förster *et al.* 2013:210–12). The caravans consist of the merchants and their escorts and pack animals, mainly donkeys/mules (see Figure 6.6) but also oxen for the journeys to the north; the total caravan consists of a minimum of half a dozen animals and sometimes more than 600.

With such a dearth of modern material on long-distance caravan operation, we can welcome Axel Nielsen's 'Ethnoarchaeological perspectives on caravan trade in the south-central Andes' (2001). This refers to llamas rather than donkeys but includes a valuable and vivid account of extensive ethnographic and ethnoarchaeological fieldwork in the Bolivian Andes, including a lengthy journey with a llama caravan in circumstances involving few 20th-century AD aids. Nielsen describes the many factors in journey decisions and scheduling, noting how crucial information was gathered from other caravaneers met *en route*.

While detailed parallels between donkey and camel caravans would be inappropriate (see Chapter 6.2), an account of late 16th-century AD camel caravans in Syria and Mesopotamia (summarised by Grant 1937:134–9) affords examples of necessary preparations for a desert journey. Security on the route was arranged with guards and through 'protection' payment to local nomadic groups. Later accounts describe how caravans were commonly large joint enterprises, for security and pooling of costs; tolls were a major burden, imposed by locals and authorities at pinch-points such as fords (Grant 1937:1127/53/166).

Early long-distance transportation

Note on the term 'trade': There is rich archaeological discussion on the definition of 'trade' as distinct from exchange, barter and other systems. Milevski (2011:7–9) prefers the Marxian term 'exchange' for transactions in periods without some form of standard equivalent, but notes that 'trade' has

Figure 6.6 Pack animals being hauled up a precipice in Africa

Source: Binger (1892:363)

an etymology relating to travel. Sherratt (1997:5) takes a robust view: '[t]he common usage of 'trade' and 'exchange', where the former is taken to imply a profit motive and the latter a socially embedded network of gifts, can easily lead to a prehistory in which there are no gainers from material transactions'. Crawford (2013:448) defines 'trade' in a Mesopotamian context as 'the commercial exchange of goods which takes place on a regular basis and where each party sees themselves as satisfied with the outcome', and for brevity and practicality I use the term 'trade' here for all such activities.

In the 6th–4th millennia BC in the Near East, regional livestock markets as still operating today could have become points of goods-exchange for seasonally mobile pastoralists. Algaze (2008:68) notes the lack of evidence for systematic long-distance human porterage in the 4th-millennium BC Near East; with Zarins (2014:188) he suggests that it was the advent of donkeys that facilitated the intensification of trade apparent in early texts. At this period, growing zooarchaeological evidence of domesticated donkeys coincides with archaeological evidence of a marked proliferation of outposts and waystations at land-water and highland-lowland intersections.

Modern and historical African examples of mobile pastoralist cultures indicate that the pack donkey has consistently been a major enabler of greater flexibility of movement, allowing groups to move further between camps (Figure 6.7), live further from water sources and exploit more marginal areas. Sherratt's (1981:287) donkey-less model of 'the added mobility given by draught and riding-animals' for 4th-millennium BC pastoralists

Figure 6.7 Tuareg child in Niger moving camp with donkey and dwelling-frames
Source: © Galen Frysinger 2001 (www.galenfrysinger.com, by permission)

was influenced by accounts of horse-riding and oxcarts in the Central Asian steppes and is now considered unlikely. Mobile groups may have become specialists in transportation of materials and goods: as an early example, the development of metallurgy east of the Zagros mountains in the 5th millennium BC, with the need to transport heavy materials long-distance in quantity, generated a new need for full-time specialist traders free of fixed pastoral itineraries, and perhaps a motivation for enlisting donkeys for pack (Matthews and Fazeli 2004:70).

Separately, there is a body of commentary asserting the primary role of river and canal transport in Mesopotamia; study of a relief map of Mesopotamia (Figure 2.1), though, convincingly argues the importance of east-west non-water routes, as evidenced by the outposts and transshipment centres developing in the mid-Uruk. As with sea, road, rail and air freight today, river- and canal-boat and pack-donkey transport systems were complementary, with the latter also providing 'feeder' local and tributary systems for the increased water traffic. Commentators on Mesopotamia may be influenced in their focus on boats by the example of Egypt, where land transportation was very much subordinate to river transport and used mainly for local carrying; but Egypt is focused on a single river and mainly lacked the major tangential land trade links long-established in Mesopotamia. Another likely influence on models is the well-known 2nd-millennium BC ruler Hammurabi's remark to the ambassadors from Zimri-Lin in a Mari text, that boats were the strength of South Mesopotamia and donkeys and wagons of the North (Archives Epistolaires de Mari, I/2, no.468 [A.1161]); this is likely to be a political over-simplification; but vehicle (though not so much donkey) use in the south may have been constrained by the marshes and canal network.

Boats had the indisputable advantage of volume (Algaze 2008:50), making them cost-efficient even when towed. A system of small rafts on the Mesopotamian rivers also existed from the 1st millennium BC to the 20th century AD, in which quffahs (circular rafts made of rawhide stretched over a basketry frame) or keleks (rafts buoyed up by inflated goatskins) travel downstream with cargo and a live donkey; on arrival, the raft/frame materials were sold and the skins transported back up the river-bank by the donkey (e.g. Admiralty War Staff 1916:166–8). Texts, though, refer to shipwrecks on both the Euphrates and the Tigris, and to seasonal low waters. The Tigris is steep, with narrow, tortuous sections and shifting sandbanks; occasionally the prevailing north-westerly wind reverses so that boats have to be towed downstream as well as up (Admiralty War Staff 1916:163/169–71). Evidence of taxation and raids affecting transport choice emerges from several of the Mari letters; donkey caravans were more able to detour to avoid taxation/customs duty posts and hostile regions but were more subject to attack, though raids were also made on boats. David Oates (1968:6) notes that

> the Great Desert route of the eighteenth and nineteenth centuries [AD] consistently kept a day's or even two days' march west of the river

> [Euphrates], since the regular toll exacted by Beduin was a lesser burden than wanton plundering by the villages of the riverain land, who enjoyed a particularly evil reputation.

Milevski (2011:195) points out, though, that the common mention of robbery of donkey caravans in the Mari texts may have more to do with 'ethnic abhorrence against nomads, mountainous population, etc' than with a genuine crime wave.

Still invisible: long-distance pack donkeys in the Ancient Near East

Until recent years there has been very little scholarly focus on the logistics of early donkey transport. Key commentators on Mesopotamia until recently made only passing reference to the donkey, and there is only very rarely speculation or interpretation on how the new donkey trains were organised and equipped, or on how and where the animals were bred, kept and disposed of. In Mesopotamian commentary the donkey is often side-lined by the traditional focus on boat transport as mentioned, while in the 3rd-millennium BC southern Levant the subject has often fallen between the stools of zooarchaeology and discussion of trade/exchange. Joan Oates (2001:279) refers to a possible caravan waystation at 3rd-millennium BC Nagar, but to date there is little evidence of the archaeological data being approached with the aim of detecting the daily practicalities of long-distance pack-donkey usage. Wilkinson (2016) argues that

> [f]or the most part, the study of large-scale trade and exchange in the ancient world remains distanced from the physical hardships of real human travel. Ancient trade or, more neutrally, ancient 'interaction' is often discussed as if it took place between actors who inhabited a flat and unchanging spatial surface. (¶1)
>
> Physical landscapes are quietly ignored in the face of social models (e.g. world-systems, peer-polity etc. are essentially social-structural models and ignore or avoid geographic determinism). (¶15)

Mesopotamia: complementary transport

There are growing signs, though, that archaeological interest is turning towards early long-distance pack-donkey transport. Algaze, in his important 2001 *Current Anthropology* piece, attributed the explosion of trade and urbanisation in the 4th millennium BC in southern Mesopotamia to 'social and "technological" innovations' (Algaze 2001:200) but referred only to water transport, dismissing donkey transport as 'less efficient' (p. 201) rather than as fulfilling a different and crucially complementary role. By 2008, though, the domestication of donkeys features in his book as a key facet of the technological basis of the advance of southern Mesopotamia

(Algaze 2008:66), with donkey caravans suggested as 'the principal method of overland carriage for long-distance trade in the area following the domestication of the donkey in the 4th millennium BC' (Algaze 2008:55). This role and its impact are more recently underlined by Zarins throughout his 2014 work *The Domestication of Equidae in Third-Millennium BCE Mesopotamia*, though his assertions suffer from his desire to ascribe similar causation and impact to the domestication of the horse in the Central Asian steppes and the donkey in the Near East, 'at the same juncture as the rise of "civilization", ca. 3000 BCE' (p. 249).

The donkey in the southern Levant: copper and Egypt

Similarly for the southern Levant, Ben-Tor in 1986 discusses trade, urbanisation and the factors in the accumulation of agricultural and other product surpluses with no mention of donkeys, but by 1992 adds the donkey and ox to his list of key factors (Ben-Tor 1992:84). Commentary on the means of land transport in this region grew further with Levy's focus on the donkey in the Negev copper trade (e.g. Levy 1995:234) – and his experimental journey with an ore-laden donkey. Copper ore was mined at Faynan from the 5th millennium BC, with some perhaps worked onsite even at this early stage; the ore and copper goods were conveyed to the Beersheva valley for working or onward supply, and this transportation process intensified throughout the 4th millennium BC and beyond. Possible donkey corrals have been found at Faynan, and domestic donkey bones and evidence of donkey-keeping at Shiqmim in the Beersheva valley. In addition to the huge quantities of charcoal (or wood for its production) required for the smelting, Mattingly *et al.* (2007:345) calculate that supplying food to the Faynan workers would require 2,800 donkey journeys per annum, plus water, pottery and personal goods. As in the region today, local pastoralists may well have been involved in the transport and trade processes; increasing use of donkeys might also be visible in greater dispersal of facilities, such as the installation of formal burying grounds well outside settlements to which the dead could now be conveyed long-distance. Camel caravans were later added as complementary to the donkey system, able to traverse the Arabah and Negev deserts but not to climb the steep sides of the plateau; mules and donkeys are used for transport in this region to this day.

There was already considerable down-the-line long-distance transportation in the 5th millennium BC between Egypt and Canaan, of heavy items such as flint, stone and shell. Prior to the systematic use of donkey caravans, these were perhaps carried by nomadic groups, although the initiators may have been Egyptian or local elites. As indicated by Sherratt's diagram (Figure 1.1), 4th-millennium BC long-distance caravans then developed, carrying olive oil, wine, and also asphalt and possibly salt. Heavier items may already also have gone by sea, but these still required feeder transport from farming and desert areas. Recent isotope analysis of a donkey skeleton from

Tell es-Safi in Israel indicates that that the animal was possibly reared in the Nile valley (Arnold *et al.* 2016).

Within Egypt, donkeys became invaluable transport animals from the 3rd millennium BC, as illustrated by Köpp (2013:110):

> [t]extual evidence shows that donkeys were used as a means of transportation on travels and expeditions, e.g. in the Old Kingdom on the journeys of Harkhuf, who took 300 donkeys with him . . . and Sabni, accompanied by 100 donkeys . . . A stela dating to the Middle Kingdom from Toshka-West mentions 1,000 donkeys accompanying a mission . . . Further missions to Sinai during the Middle Kingdom record the number of donkeys as being 200, 500, and 600.

I discuss later, in the context of physical evidence of the daily management of long-distance pack caravans, the invaluable material from the archaeological team working on the Abu Ballas trail in Egypt – a chain of staging-posts for a desert donkey-caravan route dating back to the late 3rd millennium BC (Figure 6.8), with evidence of a sophisticated system of providing 'signposts' *en route* and food and water at staging-posts (Förster 2013:297–305).

The Kaneš texts

An account of the long-distance pack donkey in the Ancient Near East would not be complete without reference to a unique later body of material,

Figure 6.8 Abu Ballas trail distinctive rock mound and storage jars

Source: Förster 2007, Fig. 13. © Rudolph Kuper; by permission

the large well-known cache of early 2nd-millennium BC clay tablets discovered at Kültepe (ancient Kaneš) in Anatolia, detailing the large-scale donkey-caravan trade in tin and textiles between here and Aššur, 1,000 caravan-kilometres to the south-east and a journey of six weeks. These describe a well-orchestrated system of transporting high-value goods using well-established transport technology: the texts indicate that this trade system was already established by c.2300 BC, though caution must of course be employed in extrapolating models back a millennium to when significant use of donkey caravans can first be deduced.

There are references in the Kaneš texts to donkeys being bought and rented *en route* and sold at the end of their journey, as the burdens on the return journey (particularly silver) were considerably less bulky; but many of such practical details of the 'donkey process' did not form part of the text-writers' concerns. Barjamovic (2011:34) speculates on the unmentioned major agricultural and logistical systems that must have arisen at waystations/inns *en route* supplying water, fodder and food for up to 300 donkeys and their drivers (and collecting dung).

There are some indications that Assyrian caravans included as many human escorts as donkeys; modern examples demonstrate that a single driver can control dozens of donkeys in a caravan, and as I note earlier in this chapter, wholly unaccompanied donkey trains are used for smuggling today worldwide; but the donkey-to-human ratio is affected on long journeys by the need for timely unloading and reloading at stops, as up to recent times with mule and reindeer pack-trains. It appears from the Kaneš tablets that the loading issue was dealt with on the 2nd-millennium BC Assyrian caravans by assigning one driver/packer (sāridum) to every one or two donkeys; this partly came about because the Kaneš caravans generally consisted of a convoy of several merchants, for support, company and security, each with his own small consignment carried on up to half a dozen donkeys, with the merchant's own drivers/packers (e.g. Dercksen 1996:255/283). Foot travellers may have also made use of the caravan to travel in safety; I refer earlier to the accounts in the 2nd-millennium BC Archives Royales de Mari of raids and organised banditry. The Kaneš caravans seem less prone to attack, perhaps through prompt reprisals or agreements with local rulers (Larsen 2015:156/176). Large caravans are only known in the later period of the Kaneš trade, perhaps because the political situation in the Syrian region had worsened by this time – a precaution taken among pack caravans to the present day.

Long-distance human porterage

Evidence of porters in antiquity

Human porters rarely appear in representations or texts in the Ancient Near East, perhaps reflecting (as with donkeys) their lowly function. It may be that much long-distance movement of materials was conducted point to point by individual mobile pastoralists and hunters, rather than by organised porterage; slaves or captives exchanged between Mesopotamian regions may also have been given packs. By the 3rd millennium BC, laden porters, alongside led equids, are shown on the Standard of Ur, and Boese (1971:168) notes representations at Tell Agrab, Khafaje and Nippur of head-loads and of a storage jar carried on a pole between two porters. As I note in Chapter 2.2, in the late 4th-millennium BC Uruk Archaic Texts captives used for labour were categorised by age and sex using the same terms as for state-owned cattle and also for imported donkeys, suggesting a perceived equivalence of human and animal labour; similar attitudes are apparent in early Indian texts (Tani 1996:403–5) and ancient Rome. The 3rd-millennium BC texts from Lagash record that sons of captive females were castrated and set to towing boats (Tani 1996:405), and daughters set to weaving, though there are accounts of females also towing boats; human towing continues in the Near East today.

Use of captives or encumbered labourers for long-distance pack carries disadvantages: they may abscond (with the goods) and will probably carry less and move more slowly than well-fed self-employed professional porters. Local porters – probably in the latter category – are mentioned in the 2nd-millennium BC Kaneš texts as being recruited within Anatolia to carry goods between towns when the pack donkeys from Aššur have been sold off. The Kaneš porters were given food and payment; there are examples of porters refusing to travel on routes with brigands or icy roads and having to be substituted by donkeys (Dercksen 1996:61–3). Human porters are 'expensive' if they are taken away from agricultural or other production, but can more easily than animals be sent away to find other employment in downtime. A separate category of travelling human in Mesopotamia comprises messengers, recorded in a few late 3rd-millennium BC texts, who travelled on foot and by boat, shifting to horseback in the 2nd millennium BC or later.

The Kaneš porters may bear some comparison with the Sherpas in modern Nepal, who are self-organised, with a system of *sirdars* (experienced foremen) to arrange groups for expeditions or large construction projects; these similarly operate a complementary system to donkey and mule caravans. In Kurdistan, both mule-caravans and *kulbar* porter groups cross the borders between Turkey, Iran and Iraq, with villagers unable to afford a mule carrying massive loads themselves.

The *tlamemes* (various spellings) of Aztec-period Mesoamerica were also professional porters, responsible for much of the long-distance transfer of raw materials and goods. *Tlamemes* were hired by Aztec merchants and normally operated on a relay system, resulting in the establishment of a chain of settlements a day's walk apart (Hassig 1985); Aztec merchants would use the same team for the whole trip only where replacements would be unavailable *en route* or local authorities were hostile. While relay systems for ridden equids, such as the post-horse system, are not uncommon in more recent times, it appears that long-distance donkey caravans, in antiquity and now, use the same animals for the whole journey. This may reflect the problems in maintaining large stocks of idle animals at each changeover point.

While the *tlamemes* were apparently male and the Sherpa porters are both male and female, portering in many parts of the world is traditionally female: in Africa, among Native Americans in pre-horse days, and among the nomads in Iran and Kurdistan. Early groups transporting items long-distance may well have included women, but evidence of any sort is slim. In Chapter 6.1 I discuss the widespread practice of short-distance female burden-carrying and the transformation of their daily lives with the adoption of donkeys for this work. Engels ([1884] 2004:65) bases part of his overall argument on the situation of women on 'barbarian' native Americans using captive women purely as wives and mothers, but ethnographic evidence indicates that they were also captured for manual labour including perhaps porterage.

Comparison of human and donkey pack capacity

There is very little consensus on the load weights that can be carried by humans; African and other figures cluster round 30–40 kg – so about half what a donkey can carry long-distance – but uniquely Sherpas in Nepal (who are small but well-muscled, with special walking and resting techniques) regularly carry well over 100 kg up the Himalayan foothills. Human loads, calculated as a percentage of body mass, are less for women who generally have less muscle per body mass. Donkeys typically carry 50–80 kg, or 100–150 kg or more for shorter distances. Calculations are complicated by algorithms of weight *vs.* distance *vs.* hours per day, the frequency of rest periods, and whether food or fodder has to be carried; but typically, humans and donkeys both walk at 4–5 kilometres/hour. Humans may cover 15–20 kilometres per day, and donkeys typically 25–30 kilometres, or sometimes up to 40.

Humans can team up to carry awkward objects (though donkeys can also be lined up to carry long items); bulk (textile, fodder) is more of an issue than weight. Humans can carry loads themselves, and lift loads over obstacles, and can potentially be sent to arrange their own food and obtain employment in gap periods. Donkeys are more energetically efficient than humans in most terrains, but humans can traverse steeper and more difficult terrain than even donkeys (see Figure 6.6). The efficiency of both is affected by size, fitness, level of nourishment, experience, muscle development related to the loading method, and whether accustomed to the climate and altitude. A key factor in the advantage of donkey pack over human is their physiological greater efficiency in transporting weight: donkeys make different and more economical use of muscles and other bodily elements for bearing loads. Yousef *et al.* (1972) suggest that these adaptations may account for donkeys' high energy output to food intake; beyond about 20% of liveweight, human energy costs for carrying rise steeply, unlike those of donkeys. The fuel cost of using donkeys, overall and per load, is thus considerably less than that for using humans; the gap is widened by donkeys' greater tolerance of heat than humans, the greater absolute load and ability to carry large-volume goods such as textiles and fodder, and the reduction of need for regular water supply and provision of food.

Another factor in choice of humans or animals for pack can be the effectiveness, in terms of design and quality, of pack equipment, which may limit the size, shape and weight of items carried by humans or by animals. The porters shown on the 3rd-millennium BC Standard of Ur seem to use a wooden pack-board, as do Nepali porters today; humans also carry items on their heads, on their backs with a head strap or as a backpack, or on poles or yokes. Finally, choices are and presumably were made on the basis of availability of humans or animals, including competitive bargaining on the part of porters and donkey caravaneers, and in the case of animals on their value at the destination.

On the road: traces of donkey-caravan operation

There is significant ongoing work on the likely route-patterns and settlement distribution of long-distance pack caravan routes in ancient Mesopotamia and elsewhere; these form very large subject areas in themselves, not to be addressed in a book on the day-to-day impact of working donkeys at ground level. The work of the Förster and Riemer team in Egypt has provided rare and welcome examples of evidence for *en-route* arrangements for donkey caravans in antiquity, on the 3rd-millennium BC Abu Ballas trail. The clear fixed route, laid out and maintained by the Pharaonic regime, contrasts with those commonly taken by caravans elsewhere avoiding attack and taxation. At the staging-posts, storage jars were replenished with water and grain by special caravans, and staff were stationed there to guard the provisions and bake bread for the caravans (Förster 2013:301–5). Similarly, in more recent Ethiopia, Helina Woldekiros (pers. comm. 2016) reports on bread-baking stones at both ancient and present-day donkey-caravan campsites.

From the distance between the staging-posts, Förster (2007:5) calculates that 'the pack animals either walked c. 40 km per day and were watered at the end of every second, or they needed three days at a rate of c. 25–30 km to cover the distance, getting their water at the end of every third'. In a more recent example, Grant (1937:63) notes that in Syria the Romans dug a well every 28 miles (45 km). Time for loading/unloading, grazing and watering is another factor affecting when and where overnight stops are made.

The Abu Ballas trail has a striking series of orientation points formed of upright stone slabs or stacked stones (Förster 2013:297; Figure 6.9). Signs of roads, for all purposes and at all periods, can include 'stones, cairns, petroglyphs, trail-marker trees, stepping stones on canoe landings, rope and

Figure 6.9 Abu Ballas trail orientation cairn and resting place
Source: Förster 2007, Fig. 6. © Rudolph Kuper, by permission

hand-and-foot rails, and wooden bridges' (Riemer and Förster 2013:29), and general evidence of investment in construction and maintenance. Clues from evidence of ritual sacrifice at the start of long journeys and at difficult crossing-points are suggested from both modern example of Bolivian llama caravans (Nielsen 2001:170–2/178–82) and accounts from the early 2nd-millennium BC Kaneš texts (Larsen 2015:177). Förster *et al.* (2013:215) also suggest an archaeobotanical role, if 'the modern distribution of certain desert plants is shaped by the animals of a caravan (through preferential consumption and spreading), and/or by humans who intentionally disseminate species to develop food resources along the travel routes'.

I have noted earlier the valuable contribution of Neilsen's (2001) report on modern llama caravans in Bolivia. He (p. 186) discusses the likely detectable features of campsites, recording that overnight camps may be more frequently placed than actual daily journey length as various factors cause delays, other caravans may already be in occupation, and caravaneers may have preferences relating to cultural group. Typically they will have a separate unloading/loading area and hearth/sleeping area, and Nielsen (2001:197) suggests that the extent of separation of sleeping-areas from the unloaded packs 'may reveal, at a behavioural level, a different relationship between drovers and the goods they transported in state-sponsored traffic' compared with goods that they own themselves.

Pack containers

On the Abu Ballas 3rd-millennium BC Egyptian desert donkey trail, the 15kg pottery water-jars found stashed along the route were likely to have been carried empty in baskets, four jars per donkey, and repeatedly filled once in place from leather water-bags carried by other donkeys. Detection of donkey transportation is, though, inevitably constrained by the near-absence in the Ancient Near Eastern archaeological record of organic material such as basketry, wood, textiles, rope and leather.

In many cultures worldwide today loads from sacks to water-drums are simply tied onto the donkey, possibly with a saddle-blanket, sack, twigs, straw or mat to protect the goods from sweat and the animal from sores. The donkey's narrow back makes paired loads most practical; these are evidenced in Near Eastern texts and representations and in modern times, and include cloth, leather, double baskets, nets, and wooden holders for stones or sacks. In particular the double cloth or woven fibre saddlebag is used widely in the Near and Middle East and Central Asia, filled with anything from flour to stones (Figure 6.10). In the Kaneš caravans, a small top-pack was often added to the paired load; top-loading may be used where the item's unit weight is too great for the animal to carry two, though with problems of stability. Water-skins, too, are sometimes slung under the animal's belly.

Milevski (2011:183–93) discusses in detail the pack containers on the well-known late 4th-millennium BC donkey figurines from the southern

Figure 6.10 Double basket

Source: © Donkey Sanctuary 2013, by permission

Levant (Figure 6.11). Although pottery containers were widely used for wine and other agricultural products, he considers (Milevski 2011:190) that

> [m]ost of the cases show an open and deep bag-shaped container. As Amiran (1985:192) has noted, the vessels in Figure 66 have no good parallels in the pottery repertoire of the period. She suggested that they are probably imitations of baskets.

It may even be that these figurines indicate not long-distance pack caravans, as habitually assumed, but local transport: the containers shown may bear closer relation to baskets for conveyance of fresh produce from the fields than to long-distance pack. Their form may of course relate more to the

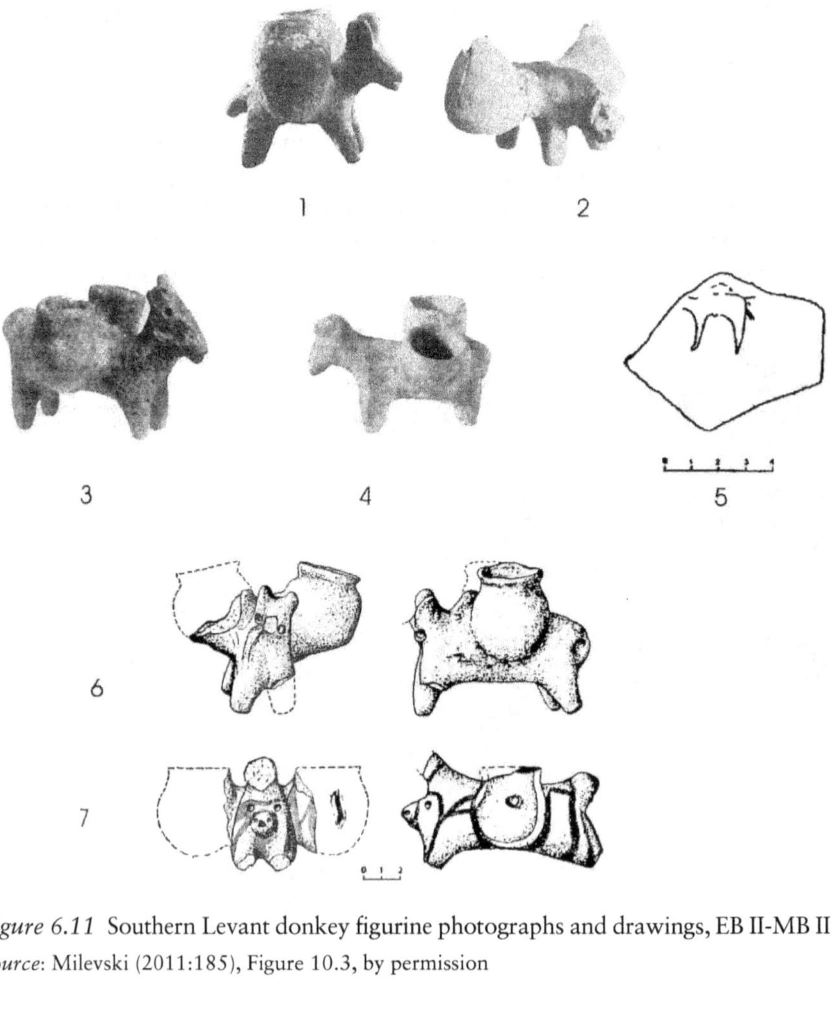

Figure 6.11 Southern Levant donkey figurine photographs and drawings, EB II-MB II

Source: Milevski (2011:185), Figure 10.3, by permission

constraints of the figurine-manufacturing process or even to possible use of the figurine for containing small funerary offerings; the risks of deducing practicalities from representations are underlined by improbable arrangements such as the well-known Chalcolithic ram figurine from Gilat, with its backload of three cornet-shaped containers.

Clues to the systematic adoption of boat transport have been sought in study of changes of jar shape, and long-distance donkey-transport might be similarly detected through changes of container shape, volume and material, including greater uniformity for speed and efficiency of loading. A key sign of ubiquity of a transport mode is its use as a standard measure: in the Kaneš texts, muttātum – half of a pair of saddlebags – was a measure for tin, and from ancient Egypt onwards, a donkey load has constituted a measure of quantity.

7 Rethinking animal ploughing

7.1 Introduction

The subject of ploughing with working animals is a vast one, encompassing not only the animals and their maintenance but also agronomic matters well beyond the scope of this book; I provide here no more than an outline of significant issues relating to practicalities that may illuminate the fragmentary working-animal evidence in Mesopotamia and beyond.

Ploughing is not necessarily an advance over manual cultivation. Ard ploughing can be deeper than manual digging, but this can be inappropriate for light, arid soils, resulting in desiccation. Ploughing opens the soil more speedily but leads to an extra labour-burden per farm of weeding and harvesting, and of preliminary field-clearing, as well as of the major task of the care and feeding of the working animals. Top-down models of the adoption of plough agriculture commonly risk 'leaving out the animals', notably factors arising from the different virtues of oxen, cows and donkeys, which can apply to large and small farms alike. Meanwhile, modern official promoters of agricultural regimes are often baffled by the lack of profit motive among smaller-scale farmers and frustrated by preferences for reduction of drudgery for family members, often achieved by use of working animals for transport rather than for ploughing.

Uptake factors

My analyses of the many development studies and overviews examined for this book demonstrate a clear hierarchy of ploughing-animal uptake factors by farmers in modern developing regions formerly using manual cultivation. Table 7.1 summarises in very broad terms the key factors put forward by commentators in more than 100 studies and overviews as affecting uptake of working animals for ploughing, by farmers formerly using manual cultivation. The factors are listed in very approximate descending order of frequency of mention in studies of animal ploughing, as a rough guide to their influence in adoption. The sources – too many to list – are largely though not wholly from modern studies (often sub-Saharan African; see Appendix B). The list and order should only be taken as a general indication of possible considerations in antiquity as they may reflect influence from modern elements such

Table 7.1 Ploughing uptake factors

Factor (approximate descending order of mention)	*Issues*
Cultural/religious barriers	Including taboos for women handling animals, cattle herders not wishing to plough
Labour	Availability of supporting labour, including for animal maintenance; use of children. Difficult to compare ancient and modern 'cost' of labour
Entry cost/return on investment	Initial step, cash flow, means of obtaining a return
Unfamiliarity with/fear of animals	Including the agriculture/herding divide
Availability of suitable animals	Obtainability of cattle and donkeys
Type of animal introduced	e.g. Cattle introduced by outside agencies where donkeys more suitable
Authorities/support	Official promotion, assistance, vetoes
Disease	Tsetse-fly regions in Africa, parasites
Training of farmers	Including learning from neighbouring groups
Market for surplus	Either demand driving production, or over-production requiring an outlet
Availability of land	Making use of spare land, or maximising use of limited land
Climate	Suitability for cattle/donkeys
Soil type	Heavy ploughing is less suitable for sandy soil
Year-round utilisation	Hiring/sharing/lending systems, use of ploughing animals for transport
Training of animals	Crucial to cost-effective ploughing
Access to grazing/fodder	Crucial for cattle in particular
Animal maintenance	Labour issues – herding, fodder-growing
Prestige cattle	Cattle kept for status
Terrain	Choice of animal; whether ploughing is an improvement over hand-cultivation
Crop type	Cash or subsistence crops, the latter perhaps linked to household manual labour
Size of household	Large household might be a driver for adopting advanced agriculture, or a deterrent as ample manual labour available
Link with transport	Animals to transport crops/implements to/from fields; ancillary farm transport; year-round use
Length of season	A short sowing 'window' may favour advanced agriculture
New farming systems needed	Changed annual cycle, e.g. sowing, weeding; need for de-stumping fields
Meat value	Cattle have residual value
Harness/equipment	Appropriate harness improves productivity

(*Continued*)

Table 7.1 (Continued)

Factor (approximate descending order of mention)	*Issues*
Theft of animals	Additionally predators; herding/corralling
Social relations	Relations between plough-users and non-users, including hiring/lending of animals and transfer of labour. Issues of communal herding
Issues for early adopters	Sometimes related to particular characteristics of early adopters
Manure	New need, for productivity; availability from working animals
Reduced burden	Eases tasks for household members – or not; increased burden of weeding
Size of farm	Large- and small-scale farmers have different strategies
Economic strategies	e.g. Buying/training/selling of working animals

as mechanised means and official/NGO intervention (though the latter is of course not only a modern phenomenon). The various factors are discussed individually throughout this book; but a key lesson for archaeologists (daunting but illuminating) is the complex interplay of factors and the failings of single-issue arguments.

'Intensive' and 'extensive': confusing terminology

There is considerable confusion in the terminology used in anthropology and archaeology for different styles of farming. The term 'intensive' is quite customarily used for large-scale centrally organised farming; Sherratt does similarly in broad discussion (1981:262) but shifts to 'extensive' (1981:292, 2006:352) in specific comparisons with 'intensive' horticultural systems; other writers too mix their appellations. Goody in 1971 (p. 25) uses 'extensive' for Africa's non-plough agriculture and 'intensive' for ploughing; in 1976 he again refers, on p. 24, to 'intensification' as relating to ploughing, but on p. 32 to male-dominated 'extensive plough agriculture', with 'intensive' applied to manual cultivation. Earlier in the same work, Goody (1976:20) uses the perhaps more secure terms 'simple' and 'advanced' for manual and plough cultivation.

Among anthropologists more widely, Boserup ([1965] 2005:43/73) defines intensification as relating to more frequent cropping as opposed to (p. 65) the 'extensive' pre-colonial system; Shenk *et al.* (2010:66) define intensification as more yield per hectare and larger fields, and Gurven *et al.* (2010:50) as associated with 'monocropping and clearcutting'. In contrast, Bogaard (e.g. 2004:21), Halstead and other archaeological commentators on agricultural prehistory relate intensive farming to small-scale mixed cultivation with high family labour input and modest use of working animals to reduce household labour rather than to increase the scale of cultivation significantly. Some anthropological commentators deal with the issue by classifying this form of intensive agriculture as horticulture and including absence of plough use in the definition (Borgerhoff Mulder *et al.* 2009:685 and others). McCown *et al.* (1979:304) go further in separating intensive horticulture from shifting agriculture as well as from extensive large-scale animal-ploughing agriculture.

Carswell (1997:1–11/21) addresses the terminology issue in detail, concluding that 'intensive' implies increase in yield per hectare whatever the cause (demand for more food, land or labour limitations) or effect. 'Extensive' implies increase in land, without an increase in labour input per hectare. Bogaard (2004:21) similarly defines extensification as involving 'smaller inputs of labour per unit area, resulting in smaller area yields'. Styring *et al.* (2017:1), in a detailed study of Mesopotamian agricultural modes, define intensive agriculture as involving an 'increase in inputs, resulting in increased crop yields per unit area of land' through high-labour practices such as manuring and weeding; agriculture based on significant expansion of cultivated land, as enabled by animal ploughing, is defined as extensive, 'such that reduction of inputs and yields per unit area are offset by a larger absolute scale of production'.

The terminology confusion may constitute part of the problem of discussing these modes of cultivation, as it is sometimes not made clear exactly what is being intensified. In this book I shall, if necessary, follow Bogaard, Halstead *et al.* in referring to small-scale high-labour farming as 'intensive' or, on occasion, follow Goody (1976:20) in using 'simple' and 'advanced'.

7.2 Ard ploughing

The earliest plough type was the ard, a simple wooden implement with a pointed pole attached to a draw-bar hitched to the animals' harness; it disturbs a line of soil while leaving the surface intact between the rows. Cross-ploughing is used if further breaking up of the soil is needed (Figure 7.1). A refinement, as described in the mid 2nd-millennium BC *Farmer's Instructions* text (Black *et al.* 1998–2006d:¶46–54), is a seeder attachment. The maresha plough widely used in Ethiopia today is an example of the ard type, often home-made from branches and sometimes without use of metal. Although the implement is simple, it can be used in a sophisticated manner to suit crops and terrain, including for weeding between crops.

The power supplied by animal traction can translate into greater ploughing depth than manual cultivation, but commentators on early animal traction risk falling into the trap of conflating deep ploughing and mouldboard ploughing. Mouldboard ploughs have a flange to invert the soil, requiring significant force; they were evolved to process effectively the damp, heavy soils of European temperate zones, burying weeds and pests too deep to survive (Brigden 1984:3). A range of on-farm studies in Africa and elsewhere, conducted by government agencies looking to provide suitable technology to regions, demonstrate that in light, dry soils this inversion instead often causes soil compaction, erosion and nutrient depletion; and indeed ard

Figure 7.1 Cross-ploughing with yoked donkeys and a maresha plough in Jawi, Ethiopia

Source: © JG Ethiopia 2014

ploughing with powerful animals can risk similar drawbacks. Light soils may need little working; Oates and Oates (1976:119) note that ard ploughing is still used in modern Mesopotamia to avoid over-cultivation, and overviews by agricultural experts report moves away from soil-inversion ploughing in some regions of sub-Saharan Africa and towards 'minimum tillage' systems, using a ripper or a simple tined point similar to an ard. In parts of Tanzania, rules were introduced after World War II banning ploughing altogether in light soils as a consequence of erosion issues.

Numbers in the plough team

The number of people and animals in a ploughing team varies significantly culturally as well as for practical reasons, with consequent effects on adoption cost and labour calculations. Comparison with evidence of team size from antiquity must also take into account categorisation practices in ancient texts: some of the animals listed as constituting a plough team in Mesopotamian texts are apparently spares, apprentices or females producing calves, while on the humans in the team, early 2nd-millennium BC Mesopotamian texts make it clear that the large 'plough teams' listed also include (sensibly for accounting purposes) the full body of workers needed to prepare the fields and deal with the harvesting and processing of crops.

Number of animals

In present-day regions using simple ploughs there are striking cultural variations – well beyond factors such as lightness of soil – as to whether a single ploughing animal or a pair (or more) is used, perhaps dating back to the harness type and farmer training process first introduced locally. A single animal is better-adapted for lighter farm tasks such as weeding, and easier to handle, but is likely to need more feeding to fuel its work. Sherratt (2006:331/344) discusses the two-animal pole/yoke system appearing in European and Near Eastern depictions from the 4th millennium BC, contrasting it with the single-animal trace-harness arrangement apparently (*pace* Littauer and Crouwel 1979:11/14) used for some contemporary sledges in the latter region, and suggesting (p. 351) that different practices in antiquity stemmed from cultural choices or other circumstances rather than from separate diffusion of discrete systems. Stol (1995:189) considers from Old Babylonian textual evidence that 'ploughs with more than two oxen were primarily used on the large estates of "public" institutions', so perhaps relegating trace harnesses as perhaps more suitable for single- or dual-animal use on smaller farms.

In Neolithic Crete, particular pathologies have been found in cattle indicating that – as commonly throughout history – each animal in a pair was habitually used only on the left or the right (Isaakidou 2008:105), deepening the investment burden as spare animals cannot easily be switched in. References in late 3rd-millennium BC Girsu texts and the early 2nd-millennium BC Code of Hammurabi also refer to oxen specially assigned to a place in the team (e.g. Renger 1990:275; Stol 1995:191). Up to eight animals in a team are mentioned in texts from the same sources, but there are indications that some may be trainee or reserve animals; or as in some regions today, animals may be rotated and changed when tired. This is also a large number to feed and care for: Heimpel (1995:97) notes that in late 3rd-millennium BC Girsu texts cultivators received extra barley for young animals, perhaps to compensate for supporting non-working animals.

In light soils, as in some regions of sub-Saharan Africa, one ploughing animal is commonly considered sufficient; even in pair-oriented Ethiopia, one animal may be used for planting, though McCann (1995:78–80) describes the elaborate systems nevertheless evolved there to assemble a pair of oxen by farmers only owning one. Using multiple animals requires particular training and more elaborate harnesses and reduces the pulling advantage (and sometimes speed) per animal but can, of course, deliver more absolute power and allow working for longer. Paired animals may plough a straighter furrow, though (especially if cattle) trample a wider area and need more turning space; the investment is higher, and more ploughing personnel may be needed (see examples to follow), though in Ethiopia and many Asian countries paired animals are guided by a single person (Blench 1997:30). The differences in performance and personnel often rest on quality of training, of the animals and the people, and of good feeding of the animals.

In developing regions today, the mixing of animal species in a team is not uncommon, with donkeys paired with mules, cows (the 82-year-old mother of an Eritrean friend ploughs with a donkey and a cow) and camels; there is less modern evidence of the combining of oxen with other species. In Mesopotamia, late 3rd-millennium BC texts list donkeys, oxen and cows as carrying out the same agricultural tasks (Heimpel 1995:91–3/134); teams were mainly solely cattle or solely donkeys, though with some temporary mixing.

Number of humans

Late 3rd-millennium BC Girsu texts describe three humans managing the plough team: one at the front for turning the team and two at the back, comprising the ploughman and an assistant (Maekawa 1984:82). The Sumerian poem *The Debate Between the Hoe and the Plough* and other contemporary texts report a six-oxen plough team as controlled by four men (Black *et al.* 1998–2006a:¶91–103); the Code of Hammurabi (§257–8) and 2nd-millennium BC texts from Mari and elsewhere indicate the plough-man at the back to be higher-status (though still lowly) than the ox-driver, who is sometimes female or a child (Stol 1995:188/192–3/207–8).

In much of modern Africa, two or three people are also commonly used, with a plougher, a leader, and an assistant (often a child or woman) with a switch to drive the pair of animals (see Figure 5.3); but in countries with a longer tradition of draught animal use such as Ethiopia and Asia, the team are commonly controlled by a single person, using voice and the touch of a whip. Renger (1990:271) takes the view that the additional individuals are employed because ploughing then becomes more accurate and efficient. Certainly it can be useful to have an individual at the head of the oxen to turn them at the end of the furrow, but Starkey (1981:13) comments that

> [s]uch excessive use of labour may even be counterproductive, for experience . . . shows that oxen walk in straighter lines and achieve better ploughing if they are controlled from behind with reins.

Paul Starkey adds (pers. comm. 2015) that in Guinea-Bissau several men are sometimes used to control and drive a two-oxen plough team in order to demonstrate to the village how wild and difficult the animals are, to lessen the chance of their being stolen; and Bwalya and Akombelwa (1999:83) report how in Zambia '[d]raft animals are regarded traditionally as wild animals which should be approached by men who seem to be stronger than women'. There may too be a subtext here echoing the prestige element evident in the ancient texts and depictions (see Figure 7.2), where men are publicly seen to be controlling fearsome animals.

A likely factor in the number needed is the level of training of both the animal and of farmers unused to close contact with draught animals. Another factor is indicated in the *Debate Between the Hoe and the Plough*, which mentions the frequent breaking-down of the plough and the need for manpower to right it (Black *et al.* 1998–2006a:¶91–103).

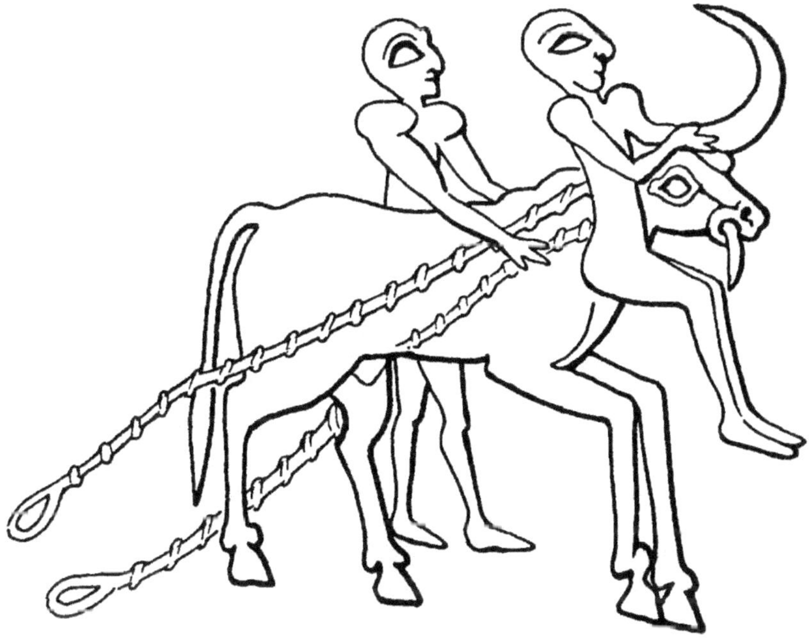

Figure 7.2 Uruk-period seal showing two men controlling a ploughing bull

Source: Sherratt (1981:264), drawn by Ann Searight. By kind permission of Ann Searight

7.3 Ploughing models and realities

The Eurasian ploughing model and Sherratt

Ester Boserup's classic 1965 model posits population growth and land shortage as drivers of the adoption of ploughing as a means of increasing yield per hectare. Isaakidou (2011:94) points out that Boserup's model 'forms the explicit basis of Sherratt's scheme for the development of farming in prehistoric Europe' from origins in the Near East, but that it is based on South-East Asian evidence and a slash-and-burn scenario inappropriate for the Near East, where 'aridity rather than forest clearance has tended to be the limiting factor'.

Boserup's early work has indeed formed a valuable grounding for further hypotheses, Sherratt's among others, on the consequences of plough adoption. Since then, though, Isaakidou (2011:97) argues, Sherratt's secondary products model has been taken up unadjusted by commentators with

> preconceived notions of technological progress and economic intensification, driven by growing human population density and an expanding

> urban world system. For those disinclined to investigate farming practices, it offered an attractively clear, off-the-peg springboard for more interesting forays into human social behaviour. The extent to which data were accommodated to the model, rather than vice versa, became increasingly clear in later expositions . . .

I firmly believe that Andrew Sherratt would have been the first to agree that his model was designed as a basis for elaboration and adjustment, not for buying off the shelf.

Jack Goody, another influential anthropological thinker, also puts in question the universal application of Eurasian-based models of the social impact of agricultural development. At farm level, while in Eurasia power and advancement may lie more in the control of land and its exploitation, he argues that in Africa (largely without the plough) land was sufficiently abundant, while shortages of labour for manual agriculture could lead to power through control of labour, including slaves. Extensive development studies in sub-Saharan Africa – in a range of environments – indicate that adoption of ploughing in light soils is likely if anything to *reduce* per-hectare yield and relies on good access to land. The modern African studies commonly describe farming communities using ploughing to extend hectarage but with low-yield practices as this incurs fewer labour costs. The assumption of increased yield nevertheless persists in academic discussion of the consequences of the introduction of deep (ox) ploughing at this period, as in for example Brewer's (2004:19) introduction to a re-publication of Engels' [1884] classic *The Origin of the Family, Private Property and the State*.

The significant land need for fodder and grazing for working animals is also often neglected in models; Boserup does refer to the subject ([1965] 2005:35–7), but in terms of additional labour for foddering rather than as a land issue. Farnham (1997:24) is one of many newer commentators underlining the effect of the new need for feeding and maintaining working animals:

> the success of animal traction in Africa appears to require abundant land for grazing and the extension of farmland. Contrary to Boserup's theory, therefore, it is most likely to succeed where population density is low.

Such discussion generally risks contrasting 'simple' ploughless horticulturalism with industrial-level plough agriculture featuring cash crops and perhaps abandonment of levelling mechanisms, whereas modern example shows a more complex array of systems, notably involving use of donkeys and cows rather than oxen. Modern African practices give ample demonstration of an intricate spectrum of immediate social effects of ploughing-animal adoption, with a substantial sector using the ard plough for family subsistence and ease of back-breaking work, with some surplus for processing and exchanging at markets for household goods. Ploughing may well be a minority usage, with donkeys in particular largely profitably used for short-distance transport, while a key element of the economics of working-animal use in history and today (but rarely included in Ancient Near Eastern models) is the widespread hiring, lending and sharing of working animals, with many modern studies of working-ox use strongly reinforcing the presence of these levelling dynamics.

Farnham (1997:35) concludes in an overview of animal traction use in modern Africa that a 'key question is whether the use of animal-drawn plows leads to socioeconomic differentiation between farmers. The available evidence is extremely murky', citing contradicting claims from a range of studies from Goody onwards. Paul Halstead has expressed doubt in recent years as to whether ox-plough agriculture in 3rd-millennium BC Greece was a major contributor to the inequality becoming apparent at the time, suggesting that it was rather a precondition that allowed investment by the better-off in draught teams, with the new asset exchangeable for labour at peak times (Halstead 2014:318). In Neolithic Crete, few signs of major inequality accompany the adoption of draught cattle; Isaakidou (2008:104–10) suggests indeed that the use of draught cattle allowed those previously disadvantaged with more distant fields to keep pace with others, and that unequal access may have been managed by community arrangements or by the use of working cows. While working cattle in the Mediterranean Neolithic were valuable not only for crop production but also in

> political economy (as means of mobilizing labor, earning prestige, and creating commensal debts) and in linking subsistence and political economy . . . [t]he durability of such inequality partly depended on the strength of leveling mechanisms.
>
> (Halstead 2014:318–19)

This echoes Bogucki's (1993:494) – and others' – view that in the European Neolithic '[m]any societies have levelling mechanisms that put a damper on excessive accumulation'.

The contribution of modern ploughing studies

The triangular calculations of area ploughed, yield, and labour use are much debated by anthropologists and agronomists, concerning plough adoption today and in antiquity. If examined in economic detail, the labour advantages of adopting ox-ploughing are not always immediately obvious. Halstead (e.g. 2014:37–9) reports that traditional farmers in the modern Aegean, their strategic farming viability proven, are often totally unable to give yield or labour figures in forms useful to agronomists. The key benefit is likely to be the greater area ploughable per season than with the hoe; but increased hectarage leads to an increased burden of manual labour for ancillary work – or perhaps not, as with small-scale farmers who prefer low yields to hiring labour for weeding.

The diagram in Figure 7.3, from a treatise on harness in modern Africa, illustrates the extreme complexity of making theoretical comparisons of working animal performance, even without additional economic and cultural factors and issues such as availability of animals and advice on their use. In practice there is a long history worldwide among farmers of pragmatic assessment of choices and trade-offs, and of empirical actions – not necessarily the theoretical best choices – taken. 'A farmer will realise without any complex analysis whether using draught animals in a particular way (or at all) is profitable or not' (Lawrence and Pearson 2002:103), though 'the economics are not always clear to the outsider' (Starkey 1994a:75, on modern African practice).

The clearest finding from a range of African and other published sources is that labour is *shifted* rather than reduced (e.g. Delgado and McIntire 1982:188): aside from weeding and harvesting, new labour tasks are created as animals require harnessing, training and maintenance as well as the feeding and watering of animals unable to graze in daytime and requiring extra nutrition for work, with concomitant land and labour requirements for fodder-growing. The labour impact of these new tasks is recognised in early 2nd-millennium BC texts from Mari and elsewhere, which indicate 'plough teams' of up to 16 individual humans but specify that some of the team members are working on harvesting, threshing, milling, reed-cutting and irrigation rather than following the plough (Stol 1995:192–5). In the slightly later Sumerian *Debate Between the Hoe and the Plough*, the entourage surrounding the plough work extensively on land preparation, weeding and de-stumping, harness-making and implement-repair (Black *et al.* 1998–2006a:¶80–90, ¶91–103).

McCann (1995:70), describing the likely consequences of the adoption of ox-ploughing in Ethiopia more than two millennia ago, reminds us that below the high-level social and economic implications there are far-reaching practical changes:

> [t]he conversion of horticultural societies, pastoralists and agrarian systems, which were based on perennial crops, to the ox-plow complex based on annual crops and integrated livestock management, implied more than a simple change of economic base or political authority. The use of the plow meant a transformation in environmental management: a livestock system which produced, trained, and maintained a stable supply of oxen, a reorganization of seasonal labor to fit a new set of crops, and a resource tenure system in which fixed multiannual resources – perennial crops – lost pride of place to annual crop fields.

Meanwhile, the information in the ancient Mesopotamian texts themselves may be suspect. Commentators have explored a range of explanations for

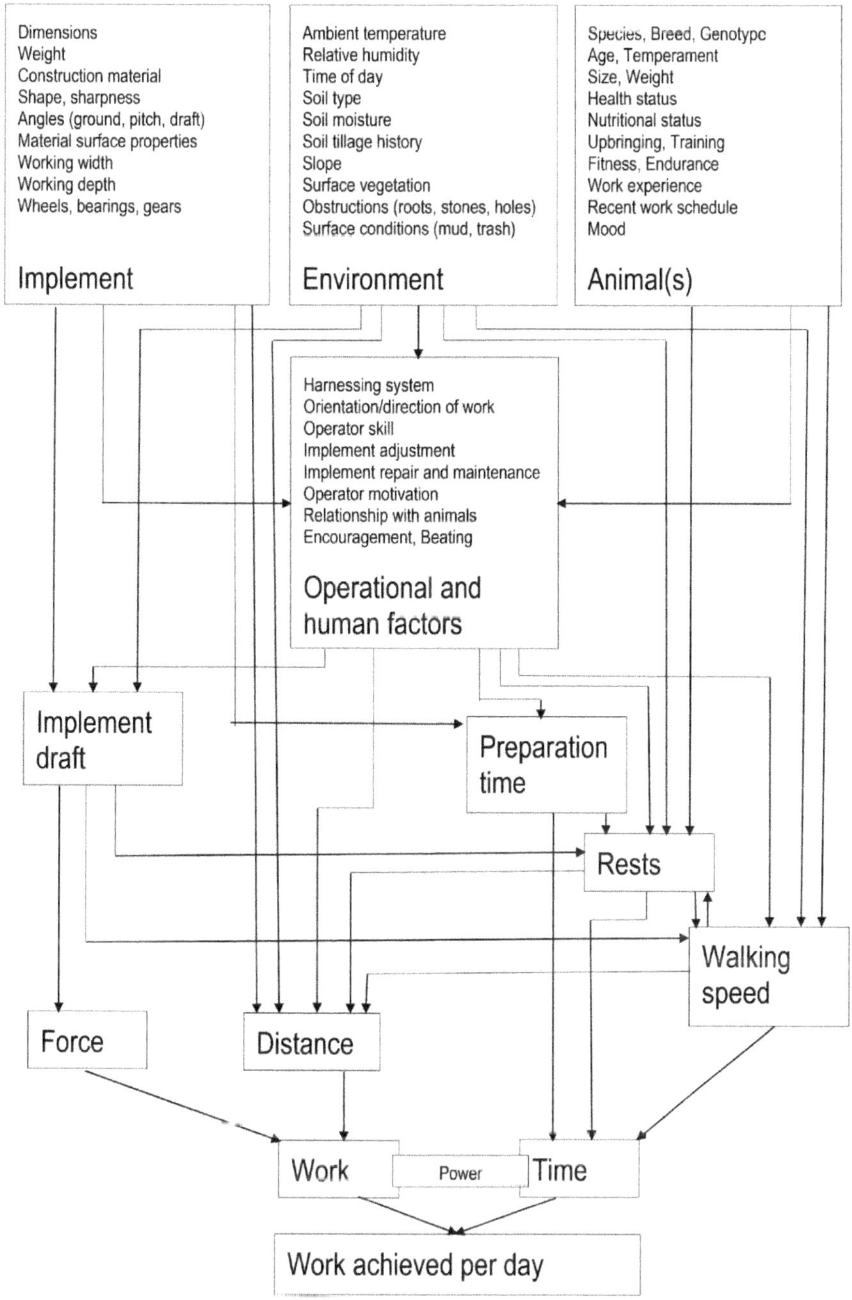

Figure 7.3 Diagram showing some influencing factors on daily work output for ploughing animals

Source: Redrawn from Starkey 1989:20, by permission

the remarkably high-yield ratios of seed to harvest in later 3rd-millennium BC Girsu and Umma texts (Maekawa 1984:82–5); Steinkeller (2004:73) argues specifically that figures in Umma texts for harvests, labour use and transactions in fact reflect 'a kind of "accounting reality" or even "accounting fiction"', as they are demonstrably largely written before the harvesting and after the projects and transactions. He suggests that these adjusted records are intended to supply planning authorities with statistical information, Liverani (2006:34) commenting that 'it was difficult to control what happened to the harvest after reaping – certainly after threshing – when it was easy to steal from it'. Adams (2008:19) proposes a very modern scenario in which the figures are perhaps adjusted by junior functionaries in order to avoid the severe penalties known to be meted out to those failing to achieve the high-performance targets set.

Year-round utilisation

Yet another factor not always included in ploughing models is the economic importance of year-round utilisation of expensive ploughing animals used only for a short season annually. This element is noted in the *Debate Between the Hoe and the Plough* (Black *et al.* 1998–2006a:¶104–8) already cited, with the Hoe accusing the Plough that 'your effective time is four months and your time of absence is eight months'. Stol (1995:184) reports on official texts on Old Babylonian cattle '[o]ne of the fears was that the oxen should remain idle (rêqum)'. Several thousand years on, government pressure in modern Greece to maximise use of family labour has led many farmers to replace cattle, taken on for ploughing, with more multi-use equids (Paul Halstead pers. comm. 2017). Jaeger and Matlon (1990:35) report on a large monitoring study of the household and practical reasons why farmers in regions of Burkina Faso have been slow to adopt animal draught power to replace manual cultivation, concluding that

> farmers can profitably adopt animal draft power when household characteristics and exogenous factors permit high utilization of animals and equipment. Empirical analysis of farm-level data indicates that low utilization is the key cause of low returns.

Sherratt in 2003 (p. 243) considers that 'the high costs of specialist draught animals' were justified by 'the use of oxen in the intensive [sic] cultivation of irrigated farmland in lowland Mesopotamia' and that other uses – threshing and eventually cart-pulling – developed in due course from this initial use. In a footnote in 2006 (p. 346) he adjusts this view, with attribution of some economic causation to non-ploughing use, though he then comes to a conclusion no longer current:

> [u]tiliser un animal de trait à la fois pour labourer/semer et pour dépiquer des céréales était un bon moyen de rentabiliser sa vie active, de

> même qu'à autres périodes de l'année son utilisation pour les transports: c'est probablement la raison pour laquelle le complexe de la traction animale a été diffusé comme un tout.

Obtaining profit from cattle in general is complex. Bogucki (1993:497) concludes in discussion of early European use of working oxen that

> [c]attle simply do not make sense as an investment for meat alone or only as insurance against agricultural deficits, particularly in an agricultural system in which the household was the primary productive unit.

An assessment of cattle in modern Asian farming systems by working-animal specialists Lawrence and Pearson (2002:104) makes the reverse point that '[d]raught work by itself is usually not profitable'. A key factor in successful adoption of working animals in 20th-century AD sub-Saharan Africa has been the additional income and general household benefit of using off-duty ploughing animals for transport, not to mention the very common hiring out (or lending for social obligation) of working animals to others. In mediaeval Britain, ox use for ploughing was contingent on additional employment for hauling carts, and in Classical Greece where 'there must exist a close connection between 'the agricultural year' and 'the construction year'' (Isager and Skydsgaard 1992:104) so that the oxen can earn their keep by transporting building materials when not ploughing. Lawrence and Pearson (2002:99) record that in Indonesia and Nepal '[m]ost animals used for draught are multi-purpose and it is often products other than their work output which determine their profitability', with the 'yield per labour unit' of owning working animals by no means stemming entirely from ploughing.

Some farmers today deal with the utilisation and cost-loss issue by selling their ploughing animals after the season, as in the example that I witnessed in Burkina Faso and as recounted to me by farmers in Ethiopia. Tiffen (1976:130) records farmers in Nigeria reluctant to adopt ploughing oxen as the promoting government forbade the sale of the oxen after the ploughing season: this is partly a training issue, as experienced farmers can rapidly train a new team every few seasons.

Ploughing animals and women

The subject of women and working animals in ancient Mesopotamia is, unfortunately, a subject whose time has not quite yet come; women high and low are largely invisible in the 4th-3rd millennium BC archaeological and textual records. While there are occasional references in Mesopotamian texts to high-status women owning cattle (e.g. Stol 1995:180, citing the 2nd-millennium BC Archives Royales de Mari (ARMT II 616:59–60) and a text from Sippar), we can assume that they did not carry out day-to-day husbandry of the animals themselves. Female slave 'ox-drivers' are mentioned in Old Babylonian texts (Stol 1995:193–4), though they are more likely to be leading the ploughing oxen (as commonly carried out by women in Africa today) than handling the plough. Again in these texts, in a plough 'team' (comprising all the workers associated with a particular plough operation) 14 men and 2 women are listed, but the women are specified as being 'millers' (Stol 1995:194). Aside from such references, we have little clue as to gender constraints on handling of working animals at any level.

The longer-term potential impacts of adoption of ox-ploughing on inheritance, marital and kinship arrangements have been well covered by anthropological commentators from Goody (1971, 1976) onwards, notably in papers in a special issue of *Current Anthropology* journal on 'Intergenerational wealth transmission and inequality in premodern societies' (2010:51/1) and in Borgerhoff Mulder *et al.* (2009:682). Engels ([1884] 2004:151) assigns the emergence of gender inequality to the advent of private property consequent upon the domestication of animals, while Brewer (2004:19–21) argues that the move from collective horticulture to individual plough agriculture led to a shift in women's work patterns. Goody (1976:33) reports from Murdock's (1967) *Ethnographic Atlas* data the correlation of 'male farming being associated with the more advanced agriculture, while female farming is found mainly with simple agriculture' with its lower potential for lucrative cash crops (Schmitz *et al.* 1991:85). Goody further (1976:25) notes an argument by some that men may have been reluctant to take on plough agriculture as it shifted the burden of cultivation to them; but as I discuss, the strong indications are that men instead use these cultural taboos to exclude women socially and to keep the power and financial advantage of ploughing to themselves.

Meanwhile, Boserup (1990:23) and others argue that plough-led advanced farming – particularly if it involves cash crops – may increase the fieldwork that therefore traditionally falls to women, who commonly take a large role in weeding and harvesting. However, Goody (1976:33) suggests that the extra 'female' work may in practice be subcontracted to cheap seasonal hired labour. A large International Livestock Research Institute study in Ghana found that use of draught oxen in practice laid more of the burden of consequent tasks on elderly and adolescent males (Panin 1987:6).

Gender-based assignment is often recorded in modern Islamic and other societies and can relate to an association throughout history and prehistory of cattle-using with manliness (see Figure 7.2 of a 4th-millennium BC seal showing two men attempting to master an ox, perhaps indicating the seal owner's ability to dominate powerful animals). Palmer (1998:144) notes that in modern northern Jordan farmers still equate the ard plough with virility and fertility, and entrenched modern local views on the unsuitability of women ploughing, generally and in particular with cattle, persist in many regions. In a wide range of modern agricultural studies of ploughing-animal use in sub-Saharan Africa and elsewhere, men are reported to consider it unsuitable for women to handle cattle, and women themselves may feel culturally or physically deterred; there can of course be other factors such as the presence of young children. In some modern societies, cattle are inherited through the male line, and even where female cattle ownership is permitted, these women may nevertheless be culturally required to hire in men to plough their land. Social analysts underline that these barriers are largely social rather than biological, with men wishing to ensure continuing labour for 'female' work such as transporting water and fuel and carrying out manual field tasks (a gender-based division already evident in late 3rd-millennium BC Mesopotamian texts), and perhaps retaining the power and financial advantage of ploughing for themselves (Goulder 2016:78).

Anthropologists hoping for models on this subject are defied by the intricate variations in sub-Saharan Africa in particular, where 'interdependent and complementary female and male farming systems exist alongside each other' (Sylwander 1994:261). Modern examples of female involvement in handling ploughing animals are not uncommon: women plough in parts of south-eastern Africa and elsewhere and take part in cattle-herding. Published material on women's use of donkeys for ploughing today is relatively scarce, though the practice certainly exists (see Figure 5.2 of an interviewee in Burkina Faso). By far the strongest working-animal story for women in modern developing regions is the transformation of their lives by the adoption of donkeys for short-distance transport.

8 Summing up
The way forward

8.1 Ox-ploughing models and the invisible donkey: lessons from modern studies

My aim has been to collate from the growing body of on-the-ground working-animal studies from Africa and elsewhere the small clues and descriptions of immediate social and economic impacts consequent on the advent of working animals in a range of modern communities, shedding light on early adaptations in the Ancient Near East. These studies make clear how choice and usage of working animals are firmly based on the physiology and behaviour of donkeys and cattle, compounded with elements such as the common cultural avoidance of eating of working donkeys. They underline the social and community adjustments – for good or bad – resulting from working-animal adoption, notably the value for small-scale farmers and for women and the disenfranchised.

The adoption of ploughing has perhaps ranked overly large in anthropological and so Mesopotamian archaeological commentary to date. Modern example shows that ploughing is not necessarily an advance over hand-cultivation and brings significant economic and labour burdens in the form of animal husbandry and year-round utilisation. Archaeological commentaries do briefly acknowledge the use of donkeys for transportation but almost exclusively in terms of long-distance pack-caravan activity. Short-distance pack work between field, farm, village and market – the bedrock of modern employment of donkeys in developing regions – is almost wholly unrecognised. Existing archaeo-anthropological models also fall short of examining the significant new activities and occupations consequent on adoption of such new technology: training, breeding, foddering, husbandry.

These new findings provide a nuanced and complex picture of local cultural adaptation to the use of working animals, clarifying the detailed effects of working-animal use which may eventually provide bottom-up adjustments to higher-level models of the role of animals in human history. In this chapter I pull out key features of my findings from my analysis of modern development studies and assessment of what they suggest to us about working animals in antiquity.

Invisible donkeys

I argue throughout that commentators on early working-animal adoption have largely relied on unquestioning acceptance of high-level European and Asian anthropological models, with "oxen" (and working animals more generally) as an abstraction rather than as living creatures with their work capabilities, maintenance and husbandry requirements. In the case of Mesopotamia there has also been undue reliance on elite-commissioned representations of oxen and ploughing in possibly ceremonial contexts. To date there has in particular been very little focus on early systematic use of working donkeys in Mesopotamia; it is often insufficiently acknowledged that these appear in texts from the late 4th millennium BC and are commonly listed as employed in centralised agricultural operations in the 3rd millennium BC; logic might also suggest the developing by-use of pack donkeys radiating from caravans traversing the region. Donkeys might usefully be regarded archaeologically in a similar light to organic materials, known to be present only by their impact on archaeologically detectable factors, with their absence from the record too readily resulting in their neglect in interpretation.

Early fieldwork in Iraq and adjoining regions was until more recent times, too, largely settlement-oriented, with the key issue that in many past and present cultures working donkeys are not eaten and their remains are rarely present in settlement food-middens. Evidence from the pioneering decades of Mesopotamian field archaeology may also inevitably lack detail and scientific rigour. These factors are exacerbated by the zooarchaeological difficulty of distinguishing between equid species, for example donkeys and the onagers native to the region and hunted for meat and hides.

Donkeys alongside cattle

Extensive modern example demonstrates that donkeys and cattle fulfil complementary working roles. Both animals can perform satisfactorily in all but extreme environments, so choice of animal (aside from artificial introduction) hangs eventually upon their physiological and behavioural differences. Cattle have greater absolute power for traction per animal, and unlike donkeys are widely used for meat at the end of their working life. Donkeys are low-maintenance: they thrive on rough forage and are widely left free to find their own sustenance; in general they are not a food source and so less susceptible to theft, and they are well able to protect themselves against predators. They have low water needs, are resistant to disease and tolerant of parasites, and are hardy and stoic. Donkeys are (despite denigratory anecdotes and traditions) easy to train and handle, quick to learn and, famously, can operate with little or no supervision, making them valuable replacements for human drudgery. They have outstanding weight-carrying ability, and a long working life subject to other factors. On the other side of the balance, donkeys in modern use may suffer from use of inappropriate harness, and they are slower than cattle to reproduce.

Small farms and donkeys – and cows

Discussion of early animal use for cultivation has often assumed the desirability of deeper ploughing as practised in temperate regions with heavy soil, and of the consequent employment of oxen. This model is countered by extensive experience in Africa, where deep ploughing can compact, erode and deplete light or arid soils; in many regions, lighter tillage, often with donkeys, is preferred, particularly on small farms. As I make clear, donkeys are widely used for ploughing in sub-Saharan Africa and other regions worldwide where mechanisation has not taken hold, particularly where grazing and water are issues and where on small farms a multi-purpose animal is desirable. While Mesopotamian texts clearly indicate massive centralised agricultural operations, the widespread presence of individual smallholdings outside the remit of the texts – perhaps using the multi-function and low-maintenance donkey – is argued increasingly by commentators, from archaeobotanical and other evidence.

One of the most striking themes emerging from modern sources is the difference in social and economic trajectories arising from adoption of working donkeys versus cattle. Wider acknowledgement of the presence and capabilities of working donkeys (and female cattle) in the Ancient Near East opens a gate to recognition of the greater complexity of working-animal systems in antiquity than was envisaged in the early days of Sherratt's secondary products model. Ploughing donkeys today are, in practice, likely to spend the majority of their time on farm or household transportation tasks, suggesting that the teams of ploughing donkeys described in 3rd-millennium BC Mesopotamian texts are very likely to have been employed much of the time for short-distance transport. This facility – unmentioned in models – provides significant social and economic agency to women and landless individuals in particular. Small-scale farmers have also regularly taken advantage throughout history of the multiple year-round utility of female cows, which have been an important ploughing animal in Egypt since ancient times and which are very widely used for work in Asia and parts of Africa.

Social levelling in the community

In an inversion of commonly held models of working animals generating wealth, the burden of feeding and husbandry of oxen in particular brings an urgent economic requirement for offsetting, through year-round utilisation outside the ploughing season. A vital contributor to this is the hiring, lending and communal ownership widespread in modern and historical working-animal systems. The circulation of working animals operates at both economic and social levels, providing currency in the form of labour or crops in return, obligation, status, and good kin and neighbour relations. Socioeconomic working-animal models nevertheless typically assume single ownership and use, though strong modern evidence suggests that models should rather be community-wide.

Little is as yet known about the role of women in use of working animals in Mesopotamia, but gender-based assignment of farming roles, notably with males and ploughing, is still strong today though by no means universal. More influential socially are the benefits brought to women by the new freedoms and social standing availed to them by the use and lending/hiring of donkeys.

Ploughing decision factors

Detailed modern evidence can also clarify confusion in theoretical discussion of the practical farming benefits or otherwise of ploughing versus manual cultivation. For example, as I describe in earlier chapters, yield increase – commonly placed at the centre of debate – is not necessarily the driver at small-farm level for adopting ploughing animals, and with the new burden of animal husbandry and feeding may shift labour rather than reducing it. Modern development studies, commonly focusing on small independent enterprises, give strong evidence of the intricacy of decision processes, by no means always driven by profit motivations.

Training of animals and users

Diffusion of new technologies is often addressed in big-picture terms, without reference to realities of how individuals learn and accept. The marked evidence from sub-Saharan African regions new to the use of working animals is that official adoption initiatives not uncommonly become subverted later, and that uptake (and choice of animal) is most successful through informal person-to-person advice and instruction on training, diffusing from village to village. A key element in successful working-animal adoption is training of users in how to get the best out of animals with least human effort; this is demonstrated by the striking regional variation in the number of humans considered necessary in a plough team despite similar circumstances.

Practicalities: breeding and foddering

Finally, modern analogy powerfully brings forward the likely proliferation of new industries and occupations consequent on the systematic adoption of working animals, which are barely discussed in models of working-animal adoption in antiquity. Mitchell (2018:3) argues that domestication of working animals is a mutual process, as humans become subjected to 'the demands that living with donkeys . . . placed on people'.

A key unconsidered activity is the mass breeding of working donkeys by specialist groups in ecologically suitable regions and their long-distance transfer (sometimes via large regional markets) to areas of high demand, as is common in modern regions of significant working-animal use. Text evidence from 3rd-millennium BC Mesopotamia supports this breeding-zone model, with potential for further identification archaeologically and zooarchaeologically. On a more local scale, feeding of working animals (particularly cattle) involves close scheduling of grazing and watering time and – given those limitations, and the additional need for nutritious food – a complex process of large-scale growing of fodder, with transportation and storage challenges. Foddering of working animals has a long history, recorded in Mesopotamian texts from the 3rd millennium BC and increasingly recognised in zooarchaeological and archaeobotanical evidence.

8.2 The way forward

I do not propose my findings as conclusions, but as incentives to revision of research approaches. The wider intention of my book is to provide a start, not an end. A key aim has been to supply a lasting compendium of information and resource material on my subject, seeking new clues to links between the animal record and the broader living contexts in which the animals are embedded, particularly through careful modern analogies and ethological consideration. In this I echo Johannsen (2011:13), who argues in his paper 'Past and present strategies for draught exploitation of draught cattle' that

> we will benefit as little from ignoring the information available on cultural variation in different types of human-animal relationships as we will from uncritically employing analogy (explicitly or implicitly).

The value of my work lies in the assembled detail, which through emerging themes and practical demonstration of working-animal systems offers new tools for archaeologists and zooarchaeologists interpreting features of Ancient Near Eastern sites. My approach is explicitly heuristic, working towards enriching the context of the higher-level anthropological models proposed by Goody and others. I argue that he and others fail to enter into the complexities of the different ploughing animals and their capabilities and needs; Goody indeed accuses himself (1976:115) of 'summarising summaries of summaries', adding that 'the view from the bird's eye does an injustice to the particular richness of life on the ground'.

A promising way forward, for reducing the invisibility of working animals and progressing towards greater appreciation of their nature, role and impact, comprises a combination of approaches including greater working-animal consciousness in excavation contexts, the pursuing of new zooarchaeological detection techniques such as isotope analysis, archaeobotanical tracing of dung concentrations, foddering and manuring, satellite imagery, and the active employment (with suitable caveats) of modern working-animal-oriented analogy to create greater understanding of the circumstances in which donkeys, oxen and cows might be systematically used for work. An ethological approach, bringing into archaeological consciousness the living working animals, places them alongside humans in the societies within which they operate. Arbuckle (2012:302) joins me in his assertion that

> Despite the fact that early texts from the late fourth and third millennia BC clearly indicate that animals and their diverse products were central concerns of early states [. . .], there have been few attempts to contextualize the role of animals within the processes that led to the rise of complex societies in the ancient Near East.

I provide material for further research into causation and circumstances, valuable not only for large-scale models of the role of working animals in 4th- and 3rd-millennium BC Mesopotamia, but also for understanding in detail the circumstances likely to cause or result in the flowering of working-animal use, and of the circumstances leading to choices between oxen, cows and donkeys. The coinciding of such suitable, desirable or exigent conditions can prove a valuable alert for site directors and laboratory researchers to work with heightened awareness of the possibility of detecting working animals when studying structures, installations, artefacts and spatial arrangements as well as animal and botanical remains. Mesopotamian archaeology in the 21st century AD is adjusting to accelerating change in fieldwork constraints, technical capabilities and interpretative challenges. New models for the developments in exploitation of secondary products in the 4th and 3rd millennia BC must now include a reassessment of the vital supporting role of the 'other' – the non-humans added to the workforce who supported social and economic transformation at the highest and lowest levels.

> A key aim of archaeological interpretation is to understand the people of the past, and it may seem banal to suggest that such people did not live in isolation in a passive, motionless world otherwise devoid of life. Humans exist within a world consisting of many other living, dynamic beings in a mutually constituting relationship. (Jones 2009:75)

The large-scale working-animal-related occupations and activities which I describe in this book are carried out by humans – the primary subjects of archaeology. While there has been acknowledgement of the importance of cattle to humans for ploughing, the working donkey in particular walks invisibly between the high-level anthropological models of Goody and Boserup with their focus on ploughing oxen, and site-focused zooarchaeological reports with their scarcity of non-food animals. The use of data and observations from widely available modern agricultural, social and economic development studies, in regions of significant working-animal use such as sub-Saharan Africa, is a largely untapped resource for Ancient Near Eastern archaeologists. With suitable caveats, it provides a means of rebooting archaeological thought and placing working animals within a newly assessed social and economic framework focusing on practicalities and on household-level responses to change.

Appendices

Appendix A
Glossary

Table B Glossary of selected terms

Reference	*Explanation*
Cattle	I refer to cattle generically, and also specifically to oxen (castrated males) and cows (this always indicates the female of the species). See Chapter 3.1 for notes on taurine and zebu cattle
Donkeys (wild and domesticated) and onagers	*Equus*, briefly for the purpose of this work, today comprises horses (*E. ferus caballus* and the related *E. ferus przewalski*), donkeys (*E. asinus*), zebras (various types including the quagga) and hemiones (*E. hemionus*, with various geographically differentiated branches including the onager, kiang and kulan, some rare or extinct). The different *Equus* species can interbreed but normally produce sterile offspring, as with the horse/donkey mule. In studies on the Ancient Near East, *E. hemionus* is commonly termed both 'hemione' and the more region-specific 'onager'; I generally use the latter term. There are some uncertain hypotheses as to the existence of a small equid, *E. hydruntinus*, possibly related to the hemione or zebra, inhabiting the mid-Levant and Anatolia but extinct since the early Holocene. *E. asinus* is descended from Nubian and (probably to a lesser extent) Somalian clades; truly wild versions are almost extinct (Geigl and Grange 2012:89). Some commentators elaborate the donkey species as the wild *E. africanus* and domesticated sub-species *E. africanus asinus*, but in view of the well-evidenced continued close blending of wild and working donkeys throughout history, I continue with the overall term *E. asinus*. Similarly, while certain commentators label non-domesticated versions as 'ass' or 'wild ass' and domesticated ones as 'donkey', I and others consider 'ass' and the more modern term 'donkey' as synonymous, and I use the latter term throughout.
Semi-wild (as in Chapters 4.3, 5.1 and 6.1)	I use this term for donkeys which have lived freely in the wild in a region over perhaps hundreds of years, long ago shedding associations with the circumstances of their original introduction, but which have become familiar over time with human groups co-living in the region, who may 'collect' them and use them *ad hoc* for transport. This term differs from 'feral', which applies to escaped domesticated animals and their immediate offspring
Trypanosomiasis	A disease affecting cattle and donkeys, endemic to parts of Africa, carried by tsetse flies

Appendix B
Examples of modern sources

For this book and its preceding PhD thesis, I have analysed in detail 389 modern studies and surveys of working-animal use in developing regions – far too many to reference and list in full in my bibliography alongside the many archaeological sources, though I include all directly referenced modern sources. I describe my data-set of modern development studies in detail in Chapter 1.2, and as an indication of the genre and context of many of my sources, I list in Table C (by date) selected key NGO, official and academic volumes, each containing multiple useful individual papers.

Many of the surveys and studies consulted come from workshop proceedings and resource books sponsored by the Animal Traction Network for Eastern and Southern Africa (ATNESA), who were very active until the early 2000s. Their website (www.atnesa.org) requires some persistence to work through, as the articles and papers published in the various volumes, some of which I mention in Table C, are mainly not listed under the volume heading; many of them are obtainable instead through a comprehensive trawl by subject and region. For researchers wishing to examine in detail as many as possible of the modern and other studies that I have consulted, in detail by subject-matter, I have created a web-page which lists the full references to all my material by the points that I have made or touched on in this book: http://www.jgoulder.com/donkeys.

Table C Some key modern development-study volumes

Type of resource	*Date*	*Title*	*Editors*	*Website/ publishers*
Workshop proceedings	1990	Animal Traction for Agricultural Development: proceedings of the 3rd Regional Workshop of the West Africa Animal Traction Network held 7–12 July, 1988, Saly, Senegal	Starkey, P. and Faye, A. (eds.)	Wageningen, Netherlands: CTA
Resource book	1991	Donkeys, Mules and Horses in Tropical Agricultural Development	Fielding, D. and Pearson, R.A. (eds.)	Edinburgh: University of Edinburgh
Workshop proceedings	1993	Research for Development of Animal Traction in West Africa. Proceedings of the fourth workshop of the West African Animal Traction Network held 9–13 July, 1990, Kano, Nigeria	Lawrence, P.R., Lawrence, K., Dijkman, J.T. and Starkey P.H. (eds.)	www.animaltraction.net/researchfor DevofATinWestAfrica.htm
Workshop proceedings	1993	Draught Animal Power in the Asian-Australasian Region. ACIAR Proceedings No. 46	Pryor, W.J. (ed.)	Canberra: Australian Centre for International Agricultural Research
Workshop proceedings	1994	Improving Animal Traction Technology: proceedings of the first workshop of the Animal Traction Network for Eastern and Southern Africa (ATNESA) held 18–23 January 1992, Lusaka, Zambia	Starkey, P., Mwenya, E. and Stares, J. (eds.)	Wageningen, Netherlands: ATNESA
Workshop proceedings	1998	Improving Donkey Utilisation and Management: report of the international ATNESA workshop held 5–9 May 1997, Debre Zeit, Ethiopia	Starkey, P. (ed.)	www.atnesa.org/atnesa-donkey-workshop-report.pdf
Resource book	1999	Conservation Tillage with Animal Traction. A resource book of the Animal Traction Network for Eastern and Southern Africa (ATNESA)	Kaumbutho, P.G. and Simalenga, T.E. (eds.)	www.atnesa.org
Resource book	1999	Meeting the Challenges of Animal Traction. A resource book of the Animal Traction Network for Eastern and Southern Africa (ATNESA), Harare, Zimbabwe	Starkey, P. and Kaumbutho, P. (eds.)	www.atnesa.org
Academic volume	2000	The Origins and Development of African Livestock: archaeology, genetics, linguistics and ethnography	Blench, R.M. and MacDonald, K.C. (eds.)	London: UCL Press

(Continued)

Table C (Continued)

Type of resource	*Date*	*Title*	*Editors*	*Website/ publishers*
Workshop proceedings	2000	Empowering Farmers with Animal Traction: proceedings of the workshop of the Animal Traction Network for Eastern and Southern Africa (ATNESA) held 20–24 September 1999, Mpumalanga, South Africa	Kaumbutho, P.G., Pearson, R.A. and Simalenga, T.E. (eds.)	www.atnesa.org/empowering1999bookcontents.htm
Workshop proceedings	2000	Animal Traction in Mozambique: a promising technology for small-scale farmers. Proceedings of the National Seminar	Mattick, A. (ed.)	www.atnesa.org/mozambique-animal-traction-workshop.pdf
Resource book	2000	Animal Power for Weed Control. A resource book of the Animal Traction Network for Eastern and Southern Africa (ATNESA)	Starkey, P. and Simalenga, T. (eds.)	www.atnesa.org
Workshop proceedings	2003	Fourth International Colloquium on Working Equines: proceedings of a colloquium organised by the Society for the Protection of Animals Abroad (SPANA) and the Faculty of Veterinary Medicine, Al Baath University, Hama, Syria, 20–26 April 2002	Pearson, R.A., Fielding, D. and Tabbaa, D. (eds.)	Wageningen, Netherlands: Wageningen Academic Publishers
Resource book	2003	Working Animals in Agriculture and Transport. A collection of some current research and development observations. EAAP Technical Series No. 6	Pearson, R.A., Lhoste, P., Saastamoinen, M. and Martin-Rosset, W. (eds.)	Wageningen, Netherlands: Wageningen Academic Publishers
Resource book	2004	Donkeys, People and Development: a resource book of the Animal Traction Network for Eastern and Southern Africa (ATNESA)	Fielding, D. and Starkey, P. (eds.)	www.atnesa.org
Workshop proceedings	2008	Animal Welfare, Livelihoods and Environment: 1st Regional Workshop, Animal Welfare Action Kenya (AWAKE) and Animal Traction Network for Eastern and Southern Africa (ATNESA), January 2008	Ochieng, F. and Wanja, A. (eds.)	Nairobi, Kenya: AWAKE, Ministry of Livestock and Fisheries Development

Workshop proceedings	2010	The 6th International Colloquium on Working Equids: learning from others. Proceedings of an International Colloquium held at the India Habitat Centre, New Delhi, India, 29th November to 2nd December 2010	The Brooke (eds.)	www.thebrooke.org/for-professionals/sixth-international-colloquium-working-equids
E-consultation	2011	Gateway to Farm Animal Welfare electronic consultation: Role, impact and welfare of working (transport and traction) animals, 1–28 February 2011	Food and Agriculture Organization of the United Nations (FAO)	www.fao.org/ag/againfo/themes/animal-welfare/blog/en
Conference abstracts	2012	The Donkey Conference, held 8–9 May 2012, SOAS, London	SOAS	www.soas.ac.uk/history/conferences/donkey-conference-2012/
Workshop proceedings	2014	Proceedings of the 7th International Colloquium on Working Equids. Royal Holloway, University of London, UK 1st-3rd July 2014	Wade, J.F. (ed.)	Norwich: World Horse Welfare

Bibliography

Adams, R.McC. (1981). *Heartland of Cities: Surveys of Ancient Settlement and Land Use on the Central Floodplain of the Euphrates*. Chicago, IL: University of Chicago Press.

Adams, R.McC. (2008). An interdisciplinary overview of a Mesopotamian city and its hinterlands. *Cuneiform Digital Library Journal* 1, §1–§11.

Adams, R.McC. (2010). Slavery and freedom in the Third Dynasty of Ur: Implications of the Garshana archives. *Cuneiform Digital Library Journal* 2, §1–§8.

Admassu, B. and Shiferaw, Y. (2011). *Donkeys, Horses and Mules: Their Contribution to People's Livelihoods in Ethiopia*. Addis Ababa, Ethiopia: The Brooke.

Admiralty War Staff (1916). *A Handbook of Mesopotamia. Vol. 1: General*. London: Admiralty War Staff Intelligence Division.

Algaze, G. (2001). Initial social complexity in Southwestern Asia: The Mesopotamian advantage. *Current Anthropology* 42, 199–233. DOI: 10.1086/320005.

Algaze, G. (2008). *Ancient Mesopotamia at the Dawn of Civilization: The Evolution of an Urban Landscape*. Chicago, IL: University of Chicago Press.

Alhaique, F. (2008). Faunal remains. In: Nigro, L. (ed.) *Khirbet al-Batrawy II: The EB II City-Gate, the EB II-III Fortifications, the EB II-III Temple: Preliminary Report of the Second (2006) and Third (2007) Seasons of Excavations*. Rome: La Sapienza, 327–58.

Anderson, P.C. (2006). Premiers tribulums, premiers tractions animales au Proche-Orient vers 8000–7500 BP? In: Pétrequin, P., Arbogast, R.-M., Pétrequin, A.-M., Van Willigen, S. and Bailly, M. (eds.) *Premiers Chariots, Premiers Araires: la diffusion de la traction animale en Europe pendant les IVe et IIIe millénaires avant notre ère*. Paris: Centre National de la Recherche Scientifique. CRA Monographies 29, 299–316.

Arbuckle, B.S. (2012). Animals and inequality in Chalcolithic central Anatolia. *Journal of Anthropological Archaeology* 31, 302–13. DOI: 10.1016/j.jaa.2012.01.008.

Archi, A. (1998). The regional state of Nagar according to the texts of Ebla. *Subartu* 4/2, 1–15.

Armstrong Oma, K. (2007). *Animals in archaeology: Between a rock and a hard place*. Paper presented at the Attitudes to Animals seminar, held 1 June 2007, Institute of Archaeology, UCL, London.

Arnold, E.R., Hartman, G., Greenfield, H.J., Shai, I., Babcock, L.E. and Maeir, A.M. (2016). Isotopic evidence for early trade in animals between Old Kingdom Egypt and Canaan. *PLoS One* 11/6, e0157650 (12 pp). DOI: 10.1371/journal.pone.0157650.

Asher-Greve, J.M. (2013). Women and agency: A survey from Late Uruk to the end of Ur III. In: Crawford, H. (ed.) *The Sumerian World*. London: Routledge, 359–77. DOI: 10.4324/9780203096604.

Audiot, A. and Garnier, J.-C. (1995). De l'Ân(e)onyme à l'hymne à l'âne ou le renversement de perspectives des usages sociaux de l'âne. *Ethnozootechnie* 56 (L'Âne), 65–77.

Ayo-Odongo, J., Mutyaba, C. and Kalunda, P. (2000). Improving on-farm transport using animal draught power in two hilly districts of Western Uganda. In: Kaumbutho, P.G., Pearson, R.A. and Simalenga, T.E. (eds.) *Empowering Farmers with Animal Traction: Proceedings of the Workshop of the Animal Traction Network for Eastern and Southern Africa (ATNESA) Held 20–24 September 1999, Mpumalanga, South Africa*. Harare, Zimbabwe: Animal Traction Network for Eastern and Southern Africa, 210–12.

Bangura, A.B. (1988). The utilization and management of draft animals at farm level. In: Starkey, P. and Ndiamé, F. (eds.) *Animal Power in Farming Systems: The Proceedings of the Second West Africa Animal Traction Networkshop, Held September 19–25, 1986, Freetown, Sierra Leone*. Braunschweig, Germany: Vieweg, for GATE/GTZ, 293–8.

Barjamovic, G. (2011). *A Historical Geography of Anatolia in the Old Assyrian Colony Period*. Copenhagen: Museum Tusculanum Press.

Barker, J. (2012). *Donkeys and mules in the 'New World'*. Conference abstract for the Donkey Conference, held 8–9 May 2012, SOAS, London.

Barrett, V., Lassiter, G., Wilcock, D., Baker, D. and Crawford, E. (1982). *Animal Traction in Eastern Upper Volta: A Technical, Economic, and Institutional Analysis*. East Lansing, MI: Department of Agricultural Economics, Michigan State University.

Beja-Pereira, A., Englund, P.R., Ferrand, N., Jordan, S., Bakhiet, A.O., Abdalla, M.A., Mashkour, M., Jordana, J., Taberlet, P. and Luikart, G. (2004). African origins of the domestic donkey. *Science* 304/5678, 1781. DOI: 10.1126/science.1096008.

Ben-Tor, A. (1986). The trade relations of Palestine in the Early Bronze Age. *Journal of the Economic and Social History of the Orient* 29, 1–27. DOI: 10.2307/3632070.

Ben-Tor, A. (1992). Early Bronze Age. In: Ben-Tor, A. (ed.) *The Archaeology of Ancient Israel*. New Haven, CT: Yale University Press, 81–125.

Binger, L.G. (1892). *Du Niger au Golfe de Guinée par le Pays de Kong et le Mossi*. Paris: Hachette.

Black, J.A., Cunningham, G., Ebeling, J., Flückiger-Hawker, E., Robson, E., Taylor, J. and Zólyomi, G. (1998–2006a). *The Debate between the Hoe and the Plough: Translation: C.5.3.1*. Oxford: The Electronic Text Corpus of Sumerian Literature.

Black, J.A., Cunningham, G., Ebeling, J., Flückiger-Hawker, E., Robson, E., Taylor, J. and Zólyomi, G. (1998–2006b). *Enmerkar and the Lord of Aratta: C.1.8.2.3*. Oxford: The Electronic Text Corpus of Sumerian Literature.

Black, J.A., Cunningham, G., Ebeling, J., Flückiger-Hawker, E., Robson, E., Taylor, J. and Zólyomi, G. (1998–2006c). *A Praise Poem of Shulgi (Shulgi A): Translation*. Oxford: The Electronic Text Corpus of Sumerian Literature.

Black, J.A., Cunningham, G., Ebeling, J., Flückiger-Hawker, E., Robson, E., Taylor, J. and Zólyomi, G. (1998–2006d). *The Farmer's Instructions: Translation*. Oxford: The Electronic Text Corpus of Sumerian Literature.

Blench, R.M. (1997). *Animal Traction in West Africa: Categories, Distribution and Constraints on Its Adoption and Further Spread: A Nigerian Case Study*. London: Overseas Development Institute. ODI Working Paper 106.

Blench, R.M. (2004). The history and spread of donkeys in Africa. In: Fielding, D. and Starkey, P. (eds.) *Donkeys, People and Development: A Resource Book of the Animal Traction Network for Eastern and Southern Africa (ATNESA)*. Wageningen, Netherlands: Technical Centre for Agricultural and Rural Cooperation (CTA), 22–30 [web edition pagination].

Blench, R.M., de Jode, A. and Gherzi, E. (2004). Donkeys in Nigeria: History, distribution and productivity. In: Fielding, D. and Starkey, P. (eds.) *Donkeys, People and Development: A Resource Book of the Animal Traction Network for Eastern and Southern Africa (ATNESA)*. Wageningen, Netherlands: Technical Centre for Agricultural and Rural Cooperation (CTA), 210–19 [web edition pagination].

Bodson, L. (1985). L'utilisation de l'âne dans l'antiquité gréco-romaine. *Ethnozootechnie* 37 (L'Âne), 7–14.

Boese, J. (1971). *Altmesopotamische Weihplatten: eine sumerische Denkmalsgattung des 3. Jahrtausends v. Chr.* Berlin: de Gruyter.

Bogaard, A. (2004). *Neolithic Farming in Central Europe: An Archaeobotanical Study of Crop Husbandry Practices*. Abingdon: Routledge. DOI: 10.4324/9780203358009.

Bogucki, P. (1993). Animal traction and household economies in Neolithic Europe. *Antiquity* 67, 492–503.

Borgerhoff Mulder, M., Bowles, S., Hertz, T., Bell, A., Beise, J., Clark, G., Fazzio, I., Gurven, M., Hill, K., Hooper, P.L., Irons, W., Kaplan, H., Leonetti, D., Low, B., Marlowe, F., McElreath, R., Naidu, S., Nolin, D., Piraino, P., Quinlan, R., Schniter, E., Sear, R., Shenk, M., Alden Smith, E., von Rueden, C. and Wiessner, P. (2009). Intergenerational wealth transmission and the dynamics of inequality in small-scale societies. *Science* 326, 682–8. DOI: 10.1126/science.1178336.

Boserup, E. ([1965] 2005). *The Conditions of Agricultural Growth: The Economics of Agrarian Change under Population Pressure*. New Brunswick, NJ: Aldine Transaction Publishers.

Boserup, E. (1990). Economic change and the roles of women. In: Tinker, I. (ed.) *Persistent Inequalities: Women and World Development*. Oxford: Oxford University Press, 14–24.

Branting, S. (2010). Agents in motion. In: Steadman, S.R. and Ross, J.C. (eds.) *Agency and Identity in the Ancient Near East: New Paths Forward*. London: Equinox, 47–59. DOI: 10.4324/9781315539256.

Brewer, D.J., Redford, D.B. and Redford, S. (1994). *Domestic Plants and Animals: The Egyptian Origins*. Warminster: Aris and Phillips.

Brewer, P. (2004). Introduction. In: Engels, F. (ed.) *The Origin of the Family, Private Property and the State [1884] (2004 Edition)*. Chippendale, NSW: Resistance Books, 7–23.

Brigden, R. (1984). *Ploughs and Ploughing*. Princes Risborough: Shire.

Brodie, N. (2008). The donkey: An appropriate technology for Early Bronze Age land transport and traction. In: Brodie, N., Doole, J., Gavalas, G. and Renfrew, C. (eds.) *Horizon: A Colloquium on the Prehistory of the Cyclades*. Cambridge: McDonald Institute for Archaeological Research, 299–304.

Bulliet, R.W. (1975). *The Camel and the Wheel*. Cambridge, MA: Harvard University Press.

Bwalya, G.M. and Akombelwa, M. (1999). Animal traction development and gender: Experiences from Western Province, Zambia. In: Starkey, P. and Kaumbutho, P. (eds.)

Meeting the Challenges of Animal Traction: A Resource Book of the Animal Traction Network for Eastern and Southern Africa (ATNESA), Harare, Zimbabwe. London: Intermediate Technology Publications, 82–5.

Cable, M. and French, F. (1950). *The Gobi Desert*. London: Readers' Union and Hodder and Stoughton.

Carswell, G. (1997). *Agricultural Intensification and Rural Sustainable Livelihoods: A 'Think Piece'*. Brighton: Institute of Development Studies. Sustainable Livelihoods Research Programme (SLP), University of Sussex. IDS Working Paper 64.

Carter, R.A. and Philip, G. (2010). Deconstructing the Ubaid. In: Carter, R.A. and Philip, G. (eds.) *Beyond the Ubaid: Transformation and Integration in the Late Prehistoric Societies of the Middle East: Papers from the Ubaid Expansion?: Cultural Meaning, Identity and the Lead-Up to Urbanism: International Workshop Held at Grey College, University of Durham, 20–22 April 2006*. Chicago, IL: Oriental Institute of the University of Chicago, 1–22.

Clarence-Smith, W.G. (2016). Equids in Mozambican history: The role of zebras, donkeys, horses, and their hybrids. *Africana Studia, Edição do Centro de Estudios Africanos da Universidade do Porto* 27, 111–25.

Clutton-Brock, J. (2003). Were the donkeys at Tell Brak (Syria) harnessed with a bit? In: Levine, M., Renfrew, C. and Boyle, K. (eds.) *Prehistoric Steppe Adaptation and the Horse*. Cambridge: McDonald Institute for Archaeological Research, 126–7.

Clutton-Brock, J. (2012). *Animals as Domesticates: A World View through History*. East Lansing, MI: Michigan State University Press.

Clutton-Brock, J. and Davies, S. (1993). More donkeys from Tell Brak. *Iraq* 55, 209–21. DOI: 10.2307/4200377.

Cochin, J. (1995). L'Ane au Sénégal et au Burkina Faso. *Ethnozootechnie* 56 (L'Âne), 55–63.

Cooper, L. (2006). *Early Urbanism on the Syrian Euphrates*. New York, NY: Routledge. DOI: 10.4324/9780203306727.

Crawford, H. (2013). Trade in the Sumerian world. In: Crawford, H. (ed.) *The Sumerian World*. London: Routledge, 447–61. DOI: 10.4324/9780203096604.

Delgado, C.L. and McIntire, J. (1982). Constraints on oxen cultivation in the Sahel. *American Journal of Agricultural Economics* 62, 188–96.

Dennis, R. and Smith, A. (1995). *Low-Cost Load-Carrying Devices: The Design and Manufacture of Some Basic Means of Transport*. London: Intermediate Technology Publications.

Dercksen, J.G. (1996). *The Old Assyrian Copper Trade in Anatolia*. Leiden: Nederlands Instituut Voor Het Nabije Oosten.

Dercksen, J.G. (2004). *The Old Assyrian Institutions*. Leiden: Nederlands Instituut Voor Het Nabije Oosten.

de Wilde, J.C. (1967). *Experiences with Agricultural Development in Tropical Africa. Vol. I: The Synthesis*. Baltimore, MD: Johns Hopkins Press.

Dietz, T., Ostrom, E. and Stern, P.C. (2003). The struggle to govern the commons. *Science* (New Series) 302/5652, 1907–12. DOI: 10.1126/science.1091015.

Dijkman, J.T. (1992). A note on the influence of negative gradients on the energy expenditure of donkeys walking, carrying and pulling loads. *Animal Production* 54, 153–6.

Dill, D.B., Yousef, M.K., Cox, C.R. and Barton, R.G. (1980). Hunger vs. thirst in the burro (*Equus asinus*). *Physiology and Behavior* 24, 975–8. DOI: 10.1016/0031-9384(80)90159-6.

Doran, J. (1994). Transportation by women, and their access to animal-drawn carts in Zimbabwe. In: Starkey, P., Mwenya, E. and Stares, J. (eds.) *Improving Animal Traction Technology: Proceedings of the First Workshop of the Animal Traction Network for Eastern and Southern Africa (ATNESA) Held 18–23 January 1992, Lusaka, Zambia*. Wageningen, Netherlands: Animal Traction Network for Eastern and Southern Africa, 272–5.

Edwards, P.R. (1983). The horse trade in Tudor and Stuart England. In: Thompson, F.M.L. (ed.) *Horses in European Economic History: A preliminary Canter*. Reading: British Agricultural History Society, 113–31.

Emberling, G. and Minc, L. (2016). Ceramics and long-distance trade in early Mesopotamian states. *Archaeological Science: Reports* 7, 819–34. DOI: 10.1016/j.jasrep.2016.02.024.

Engels, F. ([1884] 2004). *The Origin of the Family, Private Property and the State*. Chippendale, NSW: Resistance Books.

Englund, R.K. (1995). Late Uruk period cattle and dairy products: Evidence from proto-cuneiform sources. *Bulletin on Sumerian Agriculture* 8, 33–48.

Epstein, C. (1985). Laden animal figurines from the Chalcolithic period in Palestine. *Bulletin of the American Schools of Oriental Research* 258, 53–62. DOI: 10.2307/1356898.

Ewers, J.C. (1955). *The Horse in Blackfoot Indian Culture*. Washington, DC: US Government Printing Office. Smithsonian Institution Bureau of American Ethnology, Bulletin 159.

Farnham, D.A. (1997). *Plows, Prosperity and Cooperation at Agbassa: The Change from Hoes and Plows on a Government-Sponsored Land Settlement Project in Northern Togo*. Selinsgrove, PA: Susquehanna University Press.

Förster, F. (2007). With donkeys, jars and water bags into the Libyan Desert: The Abu Ballas Trail in the late Old Kingdom/First Intermediate Period. *British Museum Studies in Ancient Egypt and Sudan* 7, 1–39.

Förster, F. (2013). Beyond Dakhla: The Abu Ballas trail in the Libyan Desert. In: Förster, F. and Riemer, H. (eds.) *Desert Road Archaeology in Ancient Egypt and Beyond*. Köln: Heinrich Barth Institut, 297–337.

Förster, F., Riemer, H. and Mahir, M. (2013). Donkeys to El-Fasher or how the present informs the past. In: Förster, F. and Riemer, H. (eds.) *Desert Road Archaeology in Ancient Egypt and Beyond*. Köln: Heinrich Barth Institut, 193–218.

Gaastra, J.S., Greenfield, H.J. and Linden, V.M. (2018). Gaining traction on cattle exploitation: Zooarchaeological evidence from the Neolithic Western Balkans. *Antiquity* 92/366, 1462–77. DOI: 10.15184/aqy.2018.178.

Galvin, K.F. (1987). Forms of finance and forms of production: The evolution of specialized livestock production in the Ancient Near East. In: Brumfiel, E.M. and Earle, T.K. (eds.) *Specialization, Exchange, and Complex Societies*. Cambridge: Cambridge University Press, 119–29.

Gboku, M.L.S. (1988). Farmer social variables influencing the adopting of agricultural innovation in Sierra Leone. In: Starkey, P. and Ndiamé, F. (eds.) *Animal Power in Farming Systems: The Proceedings of the Second West Africa Animal Traction Networkshop, Held September 19–25, 1986, Freetown, Sierra Leone*. Braunschweig, Germany: Vieweg, for GATE/GTZ, 91–6.

Geigl, E.-M. and Grange, T. (2012). Eurasian wild asses in time and space: Morphological versus genetic diversity. *Annals of Anatomy* 194, 88–102. DOI: 10.1016/j.aanat.2011.06.002.

Goody, J. (1971). *Technology, Tradition, and the State in Africa*. Oxford: Oxford University Press.

Goody, J. (1976). *Production and Reproduction: A Comparative Study of the Domestic Domain*. Cambridge: Cambridge University Press.

Goulder, J. (2016). Fair exchange: Utilisation of working animals (and women) in ancient Mesopotamia and modern Africa. *Anthropology of the Middle East* 11/1, 66–84. DOI: 10.3167/ame.2016.110107.

Goulder, J. (2018). Invisible donkeys in ancient Mesopotamia: New insights from modern studies in sub-Saharan Africa. In: Çakırlar, C., Chahoud, J., Berthon, R. and Pilaar Birch, S. (eds.) *Archaeozoology of the Near East XII: Proceedings of the 12th International Symposium of the ICAZ Archaeozoology of Southwest Asia and Adjacent Areas Working Group, Groningen Institute of Archaeology, June 14–15 2015, University of Groningen, the Netherlands*. Groningen: Barkhuis Publishing and University of Groningen, 79–90.

Grant, C.P (1937). *The Syrian Desert*. London: A. and C. Black.

Greenfield, H.J., Shai, I. and Maeir, A. (2012). Being an "ass": An Early Bronze Age burial of a donkey from Tell es-Safi/Gath, Israel. *Bioarchaeology of the Near East* 6, 21–52.

Greenfield, H.J., Greenfield, T.L., Shair, I., Albaz, S. and Maeir, A.M. (2018). Household rituals and sacrificial donkeys: Why are there so many domestic donkeys buried in an Early Bronze Age neighborhood at Tell es-Safi? *Near Eastern Archaeology* 81/3, 202–11.

Grigson, C. (1993). The earliest domestic horses in the Levant? New finds from the fourth millennium of the Negev. *Journal of Archaeological Science* 20, 645–55. DOI: 10.1006/jasc.1993.1039.

Grigson, C. (2003). Animal husbandry in the late Neolithic and Chalcolithic at Arjoune: The secondary products revolution revisited. In: Parr, P.J. (ed.) *Excavations at Arjoune, Syria*. Oxford: Archaeopress, 187–240.

Groves, C.P. (1974). *Horses, Asses, and Zebras in the Wild*. London: David and Charles.

Gurven, M., Borgerhoff Mulder, M., Hooper, P.L., Kaplan, H., Quinlan, R., Sear, R., Schniter, E., von Rueden, C., Bowles, S., Hertz, T. and Bell, A. (2010). Domestication alone does not lead to inequality: Intergenerational wealth transmission among horticulturalists. *Current Anthropology* 51/1, 49–64. DOI: 10.1086/648587.

Hagmann, J. and Prasad, V.L. (1995). Use of donkeys and their draught performance in smallholder farming in Zimbabwe. *Tropical Animal Health and Production* 27/4, 231–9. DOI: 10.1007/BF02250700.

Halstead, P. (1987). Traditional and ancient rural economy in Mediterranean Europe. *Journal of Hellenic Studies* 107, 77–87. DOI: 10.2307/630071.

Halstead, P. (2014). *Two Oxen Ahead: Pre-Mechanized Farming in the Mediterranean*. Chichester: Wiley-Blackwell.

Halstead, P. and Isaakidou, V. (2011). Revolutionary secondary products: The development and significance of milking, animal-traction and wool-gathering in later prehistoric Europe and the Near East. In: Wilkinson, T.C., Sherratt, S. and Bennet, J. (eds.) *Interweaving Worlds: Systemic Interactions in Eurasia, 7th to the 1st Millennia BC*. Oxford: Oxbow, 61–76.

Hardin, G. (1968). The tragedy of the commons. *Science* 162, 1243–8. DOI: 10.1126/science.162.3859.1243.

Hart, K. (2011). *Jack Goody's Vision of World History and African Development Today: Goody Lecture 2011*. Halle, Germany: Max Planck Institute for Social Anthropology.

Hassig, R. (1985). *Trade, Tribute, and Transportation: The Sixteenth-Century Political Economy of the Valley of Mexico*. Norman, OK: University of Oklahoma Press. Civilization of the American Indian series 171.

Haufiku, N., Naunyango, S. and Rigourd, C. (2004). Donkey traction in north central Namibia. In: Fielding, D. and Starkey, P. (eds.) *Donkeys, People and Development: A Resource Book of the Animal Traction Network for Eastern and Southern Africa (ATNESA)*. Wageningen, Netherlands: Technical Centre for Agricultural and Rural Cooperation (CTA), 175–80 [web edition pagination].

Hauser, R. ([2007] 2008). *Reading Figurines: Animal Representations in Terra Cotta from Royal Building AK at Urkesh (Tell Mozan)*. Lancaster, CA: Undena Publications.

Havard, M., Vall, E., Njoya, A. and Fall, A. (2007). La traction animale en Afrique de l'Ouest et du Centre. *Travaux et Innovations* 141, 28–32.

Heimpel, W. (1994). Towards an understanding of the term siKKum. *Revue d'Assyriologie* 88, 5–31.

Heimpel, W. (1995). Plow animal inspection records from Ur III Girsu and Umma. *Bulletin on Sumerian Agriculture* 8, 71–171.

Hodder, I. (1982). *The Present Past: An Introduction to Anthropology for Archaeologists*. London: Batsford.

Ingold, T. (1980). *Hunters, Pastoralists and Ranchers: Reindeer Economies and Their Transformations*. Cambridge: Cambridge University Press. DOI: 10.1017/CBO9780511558047.

Ingold, T. (1984). Time, social relations and the exploitation of animals: Anthropological reflections on prehistory. In: Clutton-Brock, J. and Grigson, C. (eds.) *Animals and Archaeology. Vol. 3 Early Herders and Their Flocks*. Oxford: Archaeopress. British Archaeological Reports, 3–12.

Isaakidou, V. (2008). The fauna and economy of Neolithic Knossos revisited. In: Isaakidou, V. and Tomkins, P.D. (eds.) *Escaping the Labyrinth: The Cretan Neolithic in Context*. Oxford: Oxbow, 90–114.

Isaakidou, V. (2011). Gardening with cows: Hoe and plough in prehistoric Europe. In: Hadjikoumis, A., Robinson, E.N. and Viner, S. (eds.) *The Dynamics of Neolithisation in Europe: Studies in Honour of Andrew Sherratt*. Oxford: Oxbow, 90–112.

Isager, S. and Skydsgaard, J.E. (1992). *Ancient Greek Agriculture*. London: Routledge. DOI: 10.4324/9780203037744.

Jacobs, N. (2001). The Great Bophuthatswana Donkey Massacre: Discourse on the ass and politics of class and grass. *American Historical Review* 108, 485–507. DOI: 10.2307/2651615.

Jaeger, W.K. and Matlon, P.J. (1990). Utilization, profitability, and the adoption of animal draft power in West Africa. *American Journal of Agricultural Economics* 72, 35–48. DOI: 10.2307/1243143.

Jankovic, B. (2013). *Aspects of Urukean Agriculture in the First Millennium BC*. Unpublished PhD thesis, University of Vienna. http://othes.univie.ac.at/30923/1/2013-10-01_9506443.pdf [accessed 17 June 2019].

Janssen, J.J. (2005). *Donkeys at Deir el-Medina. Vol.19 Egyptologische Uitgaven*. Leiden: Nederlands Instituut voor het Nabije Oosten.

Janssen, J.J., Frood, E. and Goecke-Bauer, M. (2003). *Woodcutters, Potters and Doorkeepers: Service Personnel of the Deir el-Medina Workmen*. Leiden: Nederlands Instituut voor het Nabije Oosten.

Joffe, A. (1991). Early Bronze I and the evolution of social complexity in the Southern Levant. *Journal of Mediterranean Archaeology* 4/1, 3–58. DOI: 10.1558/jmea.v4i1.00.

Johannsen, N.N. (2005). Palaeopathology and Neolithic cattle traction: Methodological issues and archaeological perspectives. In: Davies, J., Fabis, M., Mainland, I., Richards, M. and Thomas, R. (eds.) *Diet and Health in Past Animal Populations: Current Research and Future Directions*. Oxford: Oxbow, 39–51.

Johannsen, N.N. (2011). Past and present strategies for draught exploitation of cattle. In: Albarella, U. and Trentacoste, A. (eds.) *Ethnozooarchaeology: The Present and Past of Human-Animal Relationships*. Oxford: Oxbow, 13–19.

Jones, P.A. (2008). *Major differences between work animals and food animals*. Rome: Food and Agriculture Organization of the United Nations. Paper prepared for SASAS (KZN) and DAAIG joint symposium on Understanding Communal Animal Agriculture, 30 September–2 October 2008, Drakensville ATKV Resort.

Jones, P.A. (2009). Adaptation in donkeys. *Draught Animal News* 47, 12–26.

Joubert, B. (1995). An historical perspective on animal power use in South Africa. In: Starkey, P. (ed.) *Animal Traction in South Africa: Empowering Local Communities*. Alice, South Africa: Faculty of Agriculture, University of Fort Hare, 125–53.

Kakwaba, K. and van Leeuwen, M. (1999). Hiring and lending of oxen for plowing in Kaoma, Zambia. In: Starkey, P. and Kaumbutho, P. (eds.) *Meeting the Challenges of Animal Traction: A Resource Book of the Animal Traction Network for Eastern and Southern Africa (ATNESA), Harare, Zimbabwe*. London: Intermediate Technology Publications, 308–12.

Kilemwa, A.M. (1999). Environmental impact of animal traction in Rukwa Region, Tanzania. In: Starkey, P. and Kaumbutho, P. (eds.) *Meeting the Challenges of Animal Traction: A Resource Book of the Animal Traction Network for Eastern and Southern Africa (ATNESA), Harare, Zimbabwe*. London: Intermediate Technology Publications, 68–71.

Kimura, B., Marshall, F.B., Chen, S., Rosenbom, S., Moehlman, P.D., Tuross, N., Sabin, R.C., Peters, J., Barich, B., Yohannes, H., Kebede, F., Teclai, R., Beja-Pereira, A. and Mulligan, C.J. (2010). Ancient DNA from Nubian and Somali wild ass provides insights into donkey ancestry and domestication. *Proceedings of the Royal Society of London* Series B (Biological Sciences), 1–8. DOI: 10.1098/rspb.2010.0708.

King, G. (2014). The Disperse Edaphics Project, Kenya. In: British Institute in Eastern Africa (eds.) *Annual Report 2014*. Nairobi: British Institute in Eastern Africa, 13.

Kjaerby, F. (1983). *Problems and Contradictions in the Development of Ox-Cultivation in Tanzania*. Uppsala: Scandinavian Institute of African Studies and Centre for Development Research.

Köpp, H. (2013). Desert travel and transport in ancient Egypt: An overview based on epigraphic, pictorial and archaeological evidence. In: Förster, F. and Riemer, H. (eds.) *Desert Road Archaeology in Ancient Egypt and Beyond*. Köln: Heinrich Barth Institut, 107–32.

Kramarik, J. (1975). L'histoire de l'attelage comme facteur et témoin du développement ethnique, économique et social. In: Parain, C. and Agulhon, M. (eds.) *Ethnologie et Histoire. Forces productives et problèmes de transition*. Paris: Editions Sociales, 121–36.

Kreike, E. (2010). *Deforestation and Reforestation in Namibia: The Global Consequences of Local Contradictions*. Princeton, NJ: Markus Wiener.

Kuhlmann, K.P. (2002). The "oasis bypath" or the issue of desert trade in Pharaonic times. In: Jennerstrasse 8 (eds.) *Tides of the Desert, Gezeiten der Wüste, Contributions to the Archaeology and Environmental History of Africa in Honour of Rudolph Kuper: Africa Praehistorica 14*. Köln: Heinrich Barth Institut, 125–70.

Kupper, J.-R. (1957). *Les Nomades en Mésopotamie au Temps des Rois de Mari.* Paris: Les Belles Lettres.

Landais, E. and Lhoste, P. (1990). L'association agriculture-élevage en Afrique intertropicale: un mythe techniciste confronté aux réalités du terrain. *Cahiers des Sciences Humaines* 26, 217–35.

Larsen, M.T. (2015). *Ancient Kanesh: A Merchant Colony in Bronze Age Anatolia.* New York: Cambridge University Press. DOI: 10.1017/CBO9781316344781.

Larson, G. and Fuller, D.Q. (2014). The evolution of animal domestication. *Annual Review of Ecology, Evolution, and Systematics* 45, 115–36.

Lattimore, O. (1928). *The Desert Road to Turkestan.* London: Methuen.

Lawrence, P.R. and Pearson, R.A. (2002). Use of draught animal power on small-scale mixed farms in Asia. *Agricultural Systems* 71, 99–110.

Levy, T.E. (1995). Cult, metallurgy and rank societies: The Chalcolithic period (ca.4500–3500 BCE). In: Levy, T.E. (ed.) *The Archaeology of Society in the Holy Land.* New York: Facts on File, 226–244.

Lindblom, K.G. (1931). *The Use of Oxen as Pack and Riding Animals in Africa.* Stockholm: Riksmuseets Etnografiska Avdelning.

Littauer, M.A. and Crouwel, J.H. (1979). *Wheeled Vehicles and Ridden Animals in the Ancient Near East.* Leiden: Brill.

Liverani, M. (2006). *Uruk: The First City.* Translated by Bahrani, Z. and Van de Mieroop, M. London: Equinox.

Liverani, M. and Heimpel, W. (1995). Observations on livestock management in Babylonia. *Acta Sumerologica* 17, 127–44.

Maekawa, K. (1979). Animal and human castration in Sumer part I: Cattle (gu4) and equids (ANŠE.DUN.GI, ANŠE.BARxAN) in pre-Sargonic Lagash. *Zinbun: Memoirs of the Research Institute for Humanistic Studies* 15, 95–137.

Maekawa, K. (1984). Cereal cultivation in the Ur III period. *Bulletin on Sumerian Agriculture* 1, 73–96.

Makwanda, A., Shemdoe, M.S. and Msagusa, M. (2000). Experience in the promotion of animal-powered weeding in Tanga Region, Tanzania. In: Starkey, P. and Simalenga, T. (eds.) *Animal Power for Weed Control: A Resource Book of the Animal Traction Network for Eastern and Southern Africa (ATNESA).* Wageningen, The Netherlands: Technical Centre for Agricultural and Rural Cooperation (CTA), 173–6.

Marshall, F. (2007). African pastoral perspectives on domestication of the donkey: A first synthesis. In: Denham, T., Iriarte, J. and Vrydaghs, L. (eds.) *Rethinking Agriculture: Archaeological and Ethnoarchaeological Perspectives.* Walnut Creek, CA: Left Coast Press, 371–407.

Marshall, F., Dobney, K., Denham, T. and Capriles, J.M. (2014). Evaluating the roles of directed breeding and gene flow in animal domestication. *Proceedings of the National Academy of Sciences of the United States of America* 111/17, 6153–8. DOI: 10.1073/pnas.1312984110.

Marshall, F. and Weissbrod, L. (2009). The consequences of women's use of donkeys for pastoral flexibility: Maasai ethnoarchaeology. In: Grupe, G., McGlynn, G. and Peters, J. (eds.) *Tracking Down the Past: Ethnohistory Meets Archaeozoology: Documenta Archaeobiologiae no. 7.* Rahden, Germany: Marie Leidorf, 59–79.

Marshall, F. and Weissbrod, L. (2011). Domestication processes and morphological change: Through the lens of the donkey and African pastoralism. *Current Anthropology* 52/S4, S317–S413. DOI: 10.1086/658389.

Marshall, K. and Starkey, P. (1998). Donkeys and development: Socio-economic issues: Points from keynote presentation of K Marshall and P Starkey, drawing on the lead paper of P. Fernando, G. Bwalya, S. Croxton, K. Marshall, M. Moorosi, M. Mrema and P. Starkey. In: Starkey, P. (ed.) *Improving Donkey Utilisation and Management: Report of the International ATNESA Workshop Held 5–9 May 1997, Debre Zeit, Ethiopia*. Harare, Zimbabwe: Animal Traction Network for Eastern and Southern Africa (ATNESA), 33–4.

Matthewman, R.W. (1987). Role and potential of draught cows in tropical farming systems: A review. *Tropical Animal Health and Production* 19, 215–22. DOI: 10.1007/BF02242119.

Matthews, R. and Fazeli, H. (2004). Copper and complexity: Iran and Mesopotamia in the fourth millennium B.C. *Iran* 42, 61–75. DOI: 10.2307/4300663.

Matthews, V.H. (1978). *Pastoral Nomadism in the Mari Kingdom (ca. 1830–1760 B.C.)*. Cambridge, MA: American Schools of Oriental Research.

Mattingly, D., Newson, P., Creighton, O., Tomber, R., Grattan, J., Hunt, C., Gilbertson, D., el-Rishi, H. and Pyatt, B. (2007). A landscape of imperial power: Roman and Byzantine Phaino. In: Barker, G., Gilbertson, D. and Mattingly, D. (eds.) *Archaeology and Desertification: The Degradation and Well-Being of the Wadi Faynan Landscape, Southern Jordan*. Oxford: Oxbow Books, 305–48.

McCann, J. (1984). *Plows, Oxen, and Household Managers: A Reconsideration of the Land Paradigm and the Production Equation in Northeast Ethiopia*. Boston, MA: African Studies Center, Boston University. Working Paper 95.

McCann, J. (1995). *People of the Plow: An Agricultural History of Ethiopia, 1800–1990*. Madison, WI: University of Wisconsin Press.

McCown, R.L., Haaland, G. and de Haan, C. (1979). The interaction between cultivation and livestock production in semi-arid Africa. *Ecological Studies* 34, 297–332. DOI: 10.1007/978-3-642-67328-3_13.

McGreevy, P. (2004). *Equine Behavior: A Guide for Veterinarians and Equine Scientists*. Edinburgh: Saunders.

McMahon, A. (2013). North Mesopotamia in the third millennium BC. In: Crawford, H. (ed.) *The Sumerian World*. London: Routledge, 462–77. DOI: 10.4324/9780203096604.

McShane, C. and Tarr, J.A. (2006). The horse as technology: The city animal as cyborg. In: Olsen, S.L., Grant, S., Choyke, A.M. and Bartosiewicz, L. (eds.) *Horses and Humans: The Evolution of Human-Equine Relationships*. Oxford: Archaeopress. British Archaeological Reports, International Series 1560, 365–75.

McShane, C. and Tarr, J.A. (2007). *The Horse in the City: Living Machines in the Nineteenth Century*. Baltimore, MD: The Johns Hopkins University Press.

Meadow, R.H. (1984). Animal domestication in the Middle East: A view from the eastern margin. In: Clutton-Brock, J. and Grigson, C. (eds.) *Animals and Archaeology. Vol. 3: Early Herders and Their Flocks*. Oxford: Archaeopress. British Archaeological Reports, 309–37.

Mgabo, M.R. and Masune, K.M. (2012). Utilization of draught animals for improving agriculture production in Dodoma region: A case of Msisi ward. *International Journal of Physical and Social Sciences* 2/8, 179–98.

Michel, C. (2004). The perdum-mule, a mount for distinguished persons in Mesopotamia during the first half of the second millennium BC. In: Santillo Frizell, B. (ed.) *Man and Animal in Antiquity: Proceedings of the Conference at the Swedish Institute in Rome, September 9–12, 2002*. Rome: Swedish Institute in Rome, 190–200.

Milevski, I. (2011). *Early Bronze Age Goods Exchange in the Southern Levant: A Marxist Perspective*. London: Equinox Publishing.

Mitchell, P. (2018). *The Donkey in Human Prehistory: An Archaeological Perspective*. Oxford: Oxford University Press.

Münzinger, P. (1982). Development and situation of animal traction in Africa and conditions limiting further promotion. In: Münzinger, P. (ed.) *Animal Traction in Africa*. Eschborn, Germany: Vieweg, for GATE/GTZ, 15–61.

Murdock, G.P. (1967). *Ethnographic Atlas*. Pittsburgh, PA: University of Pittsburgh Press.

Mutemi, M. (2008). Trading in donkeys. In: Ochieng, F. and Wanja, A. (eds.) *Animal Welfare, Livelihoods and Environment: 1st Regional Workshop, Animal Welfare Action Kenya (AWAKE) and Animal Traction Network for Eastern and Southern Africa (ATNESA), January 2008*. Nairobi, Kenya: Animal Welfare Action Kenya (AWAKE), Ministry of Livestock and Fisheries Development, 69.

Mutua, J. (2004). Some challenges to the use of donkeys in Kenya. In: Fielding, D. and Starkey, P. (eds.) *Donkeys, People and Development: A Resource Book of the Animal Traction Network for Eastern and Southern Africa (ATNESA)*. Wageningen, Netherlands: Technical Centre for Agricultural and Rural Cooperation (CTA), 99–101 [web edition pagination].

Nicolaisen, J. (1963). *Ecology and Culture of the Pastoral Tuareg*. Copenhagen: The National Museum of Copenhagen.

Nielsen, A.E. (2001). Ethnoarchaeological perspectives on caravan trade in the South-Central Andes. In: Kuznar, L.A. (ed.) *Ethnoarchaeology of Andean South America: Contributions to Archaeological Method and Theory*. Ann Arbor, MI: International Monographs in Prehistory, 163–201.

Oates, D. (1968). *Studies in the Ancient History of Northern Iraq*. London: Oxford University Press.

Oates, D. and Oates, J. (1976). Early irrigation agriculture in Mesopotamia. In: Sieveking, G. de G., Longworth, I.H. and Wilson, K.E. (eds.) *Problems in Economic and Social Archaeology*. London: Duckworth, 109–35.

Oates, D. and Oates, J. (2001). The excavations. In: Oates, D., Oates, J. and McDonald, H. (eds.) *Excavations at Tell Brak. Vol. 2: Report on the 3rd Millennium B.C. Material*. Cambridge: McDonald Institute for Archaeological Research; British School of Archaeology in Iraq, 15–98.

Oates, J. (2001). Equid figurines and 'chariot' models. In: Oates, D., Oates, J. and McDonald, H. (eds.) *Excavations at Tell Brak. Vol. 2: Report on the 3rd Millennium B.C. Material*. Cambridge: McDonald Institute for Archaeological Research: British School of Archaeology in Iraq, 279–83.

Oren, E. (1989). Early Bronze Age settlement in northern Sinai: A model for Egypto-Caananite interconnections. In: de Miroschedji, P. (ed.) *L'Urbanisation de la Palestine à l'Age du Bronze Ancien*. Oxford: Archaeopress. British Archaeological Reports, 389–405.

Osborn, D.J. and Osbornová, J. (1998). *The Mammals of Ancient Egypt*, Vol. 4. Warminster: Aris and Phillips. Natural History of Egypt.

Ostrom, E., Burger, J., Field, C.B., Norgaard, R.B. and Policansky, D. (1999). Revisiting the commons: Local lessons, global challenges. *Science* (New Series) 284/5412, 278–82. DOI: 10.1126/science.284.5412.278.

Ovadia, E. (1992). The domestication of the ass and pack transport by animals: A case of technological change. In: Bar-Yosef, O. and Khazanov, A. (eds.) *Pastoralism in the Levant: Archaeological Materials in Anthropological Perspectives*. Madison, WI: Prehistory Press, 19–28.

Palmer, C. (1998). Following the plough. *Levant* 30, 129–65.

Palmer, C., Smith, H. and Daly, P. (2007). Ethnoarchaeology. In: Barker, G., Gilbertson, D.D. and Mattingly, D.J. (eds.) *Archaeology and Desertification: The Degradation and Well-Being of the Wadi Faynan Landscape, Southern Jordan*. Oxford: Oxbow, 369–96.

Panin, A. (1987). The use of bullock traction technology for crop cultivation in northern Ghana: An empirical economic analysis. *ILCA Bulletin [International Livestock Research Institute]* 29, 2–8.

Panin, A. and Ellis-Jones, J. (1994). Increasing the profitability of draft animal power. In: Starkey, P., Mwenya, E. and Stares, J. (eds.) *Improving Animal Traction Technology: Proceedings of the First Workshop of the Animal Traction Network for Eastern and Southern Africa (ATNESA) Held 18–23 January 1992, Lusaka, Zambia*. Wageningen, Netherlands: Animal Traction Network for Eastern and Southern Africa, 94–103.

Pankhurst, R. (1968). *Economic History of Ethiopia, 1800–1935*. Addis Ababa: Haile Selassie I University Press.

Payne, S. (1988). Animal bones from Tell Rubeidheh. In: Killick, R.G. (ed.) *Tell Rubeidheh: An Uruk Village in the Jebel Hamrin: By T. Cuyler Young Jr*. Warminster: British School of Archaeology in Iraq and the Directorate of Antiquities; Aris and Phillips, 98–135.

Pearson, R.A. (1998). Improving the management of donkeys in Africa: Points from the keynote presentation by R.A. Pearson based on the paper prepared by R.A. Pearson, E.M. Nengomasha and R.C. Krecek. In: Starkey, P. (ed.) *Improving Donkey Utilisation and Management: Report of the International ATNESA Workshop Held 5–9 May 1997, Debre Zeit, Ethiopia*. Harare, Zimbabwe: Animal Traction Network for Eastern and Southern Africa (ATNESA), 24–5.

Pearson, R.A., Alemayehu, M., Tesfaye, A., Allan, E.F., Smith, D.G. and Asfaw, M. (2001). *Use and Management of Donkeys in Peri-Urban Areas of Ethiopia: Report of Phase One of the CTVM/EARO Collaborative Project, April 1999–June 2000*. Roslin: Centre for Tropical Veterinary Medicine, Draught Animal Power Technical Report 5.

Pearson, R.A. and Vall, E. (1998). Performance and management of draft animals in agriculture in sub-Saharan Africa: A review. *Tropical Animal Health and Production* 30/5, 309–24. DOI: 10.1023/A:1005059308088.

Peters, P.E. (1986). Household management in Botswana: Cattle, crops, and wage labor. In: Moock, J.L. (ed.) *Understanding Africa's Rural Households and Farming Systems*. Boulder, CO: Westview, 133–54.

Pétrequin, P., Arbogast, R.-M., Pétrequin, A.-M., Van Willigen, S. and Bailly, M. (2006). La traction animale au Néolithique: diversité des documents, diversité des approches. In: Pétrequin, P., Arbogast, R.-M., Pétrequin, A.-M., Van Willigen, S. and Bailly, M. (eds.) *Premiers Chariots, Premiers Araires: la diffusion de la traction animale en Europe pendant les IVe et IIIe millénaires avant notre ère*. Paris: Centre National de la Recherche Scientifique, CRA Monographies 29, 11–20.

Pettinato, G. (1991). *Ebla: A New Look at History*. Baltimore, MD: Johns Hopkins University.

Piggott, S. (1983). *The Earliest Wheeled Transport: From the Atlantic Coast to the Caspian Sea*. London: Thames and Hudson.

Pingali, P., Bigot, Y. and Binswanger, H. (1987). *Agricultural Mechanization and the Evolution of Farming Systems in Sub-Saharan Africa*. Washington, DC: World Bank and Johns Hopkins University Press.

Pollock, S. (1999). *Ancient Mesopotamia*. Cambridge: Cambridge University Press.

Porter, A. (2012). *Mobile Pastoralism and the Formation of Near Eastern Civilizations: Weaving Together Society*. Cambridge: Cambridge University Press. DOI: 10.1017/CBO9780511895012.

Postgate, J.N. (1986). The equids of Sumer, again. In: Meadow, R.H. and Uerpmann, H.-P. (eds.) *Equids in the Ancient World*, Vol. 1. Wiesbaden: Reichert, 194–206.

Potts, D.T. (2011). *Equus asinus* in highland Iran: Evidence old and new. In: Conard, N.J., Drechsler, P. and Morales, A. (eds.) *Between Sand and Sea: The Archaeology and Human Ecology of Southwestern Asia: Festschrift in Honor of Hans-Peter Uerpmann*. Tübingen: Kerns Verlag, 167–75.

Pournelle, J.R. (2003). *Marshland of Cities: Deltaic Landscapes and the Evolution of Early Mesopotamian Civilization*. Unpublished Ph.D. thesis, University of California, San Diego, CA [accessed 17 June 2019].

Pournelle, J.R. and Algaze, G. (2014). Travels in Edin: Deltaic resilience and early urbanism in greater Mesopotamia. In: McMahon, A. and Crawford, H.E.W. (eds.) *Preludes to Urbanism: The Late Chalcolithic of Mesopotamia: In Honour of Joan Oates*. Cambridge: McDonald Institute for Archaeological Research, 7–34.

Power, R.K. (2004). Deconstructing, deciphering and dating the donkey in Old Kingdom wall painting and relief. *The Bulletin of the Australian Centre for Egyptology* 15, 131–51.

Prasad, V.L., Marovanidze, K. and Nyathi, P. (1991). The use of donkeys as draught animals relative to bovines in the communal farming sector of Zimbabwe. In: Fielding, D. and Pearson, R.A. (eds.) *Donkeys, Mules and Horses in Tropical Agricultural Development*. Edinburgh: University of Edinburgh, 231–9.

Redding, R.W. (1992). Egyptian Old Kingdom patterns of animal use and the value of faunal data in modeling socioeconomic systems. *Paléorient* 18/2, 99–107.

Reh, I. (1982). Animal husbandry and veterinary aspects. In: Münzinger, P. (ed.) *Animal Traction in Africa*. Eschborn, Germany: Vieweg, for GATE/GTZ, 67–132.

Renger, J. (1990). Report on the implications of employing draught animals. *Bulletin on Sumerian Agriculture* 5, 267–79.

Riemer, H. and Förster, F. (2013). Ancient desert roads: Towards establishing a new field of archaeological research. In: Förster, F. and Riemer, H. (eds.) *Desert Road Archaeology in Ancient Egypt and Beyond*. Köln: Heinrich Barth Institut, 19–58.

Robertshaw, P.T. (1978). The archaeology of an abandoned pastoralist camp-site. *South African Journal of Science* 74, 29–31.

Robertshaw, P.T. and Collett, D.P. (1983). The identification of pastoral peoples in the archaeological record: An example from East Africa. *World Archaeology* 15/1, 67–78. DOI: 10.1080/00438243.1983.9979885.

Robinson, R. (1977). *Ways to Move: The Geography of Networks and Accessibility*. Cambridge: Cambridge University Press.

Roth, M. (1980). The scholastic exercise 'Laws about rented oxen'. *Journal of Cuneiform Studies* 32/4, 127–46. DOI: 10.2307/1359651.

Russell, N. (1998). Cattle as wealth in Neolithic Europe: Where's the beef? In: Bailey, D.W. (ed.) *The Archaeology of Value: Essays on Prestige and the Processes of Valuation*. Oxford: Archaeopress, 42–54.

Russell, N. (2012). *Social Zooarchaeology: Humans and Animals in Prehistory*. Cambridge: Cambridge University Press. DOI: 10.1017/CBO9781139019712.

Sallaberger, W. (2014). Urban organizations for offerings, overland traffic and the Euphrates trade at pre-Sargonic Mari. In: Butterlin, P., Margueron, J.-C., Muller,

B., Al-Maqdissi, M., Beyer, D. and Cavigneaux, A. (eds.) *Mari, ni Est, ni Ouest: actes du colloque "Mari, ni Est ni Ouest" tenu les 20–22 octobre 2010 à Damas, Syrie. Supplément Syria II, Vol. I.* Beyrouth: Presses de l'Ifpo, 341–54.

Sasson, A. (2010). *Animal Husbandry in Ancient Israel: A Zooarchaeological Perspective on Livestock Exploitation, Herd Management and Economic Strategies.* London: Equinox Publishing.

Schmitz, H., Sommer, M. and Walter, S. (1991). *Animal Traction in Rainfed Agriculture in Africa and South America.* Braunschweig, Germany: Vieweg, for GATE.

Shackelford, L., Marshall, F. and Peters, J. (2013). Identifying donkey domestication through changes in cross-sectional geometry of long bones. *Journal of Archaeological Science* 40, 4170–79. DOI: 10.1016/j.jas.2013.06.006.

Shahack-Gross, R., Simons, A. and Ambrose, S.H. (2008). Identification of pastoral sites using stable nitrogen and carbon isotopes from bulk sediment samples: A case study in modern and archaeological pastoral settlements in Kenya. *Journal of Archaeological Science* 30, 439–59. DOI: 10.1016/j.jas.2007.07.002.

Shai, I., Greenfield, H.J., Brown, A., Albaz, S. and Maeir, A.M. (2016). The importance of the donkey as a pack animal in the Early Bronze Age southern Levant: A view from Tell es-Safi/Gath. *Zeitschrift des Deutschen Palastina-Vereins* 132/1, 1–25.

Shenk, M., Borgerhoff Mulder, M., Beise, J., Clark, G., Irons, W., Leonetti, D., Low, B.S., Bowles, S., Hertz, T., Bell, A. and Piraino, P. (2010). Intergenerational wealth transmission among agriculturalists: Foundations of agrarian inequality. *Current Anthropology* 51/1, 65–83. DOI: 10.1086/648658.

Shennan, S. (2004). Analytical archaeology. In: Bintliff, J. (ed.) *A Companion to Archaeology*. Oxford: Blackwell, 3–20.

Sherratt, A.G. (1981). Plough and pastoralism: Aspects of the secondary products revolution. In: Hodder, I., Isaac, G. and Hammond, N. (eds.) *Pattern of the Past: Studies in Honour of David Clarke.* Cambridge: Cambridge University Press, 261–305.

Sherratt, A.G. (1997). Introduction: Changing perspectives on European prehistory. In: Sherratt, A.G. (ed.) *Economy and Society in Prehistoric Europe: Changing Perspectives.* Princeton, NJ: Princeton University Press, 1–34.

Sherratt, A.G. (2003). The horse and the wheel: The dialectics of change in the circum-Pontic region and adjacent areas. In: Levine, M., Renfrew, C. and Boyle, K. (eds.) *Prehistoric Steppe Adaptation and the Horse.* Cambridge: McDonald Institute Monographs, 233–52.

Sherratt, A.G. (2006). La traction animale et la transformation de l'Europe néolithique. In: Pétrequin, P., Arbogast, R.-M., Pétrequin, A.-M., Van Willigen, S. and Bailly, M. (eds.) *Premiers Chariots, Premiers Araires: la diffusion de la traction animale en Europe pendant les IVe et IIIe millénaires avant notre ère.* Paris: Centre National de la Recherche Scientifique. CRA Monographies 29, 329–60.

Singh, R.D. (1988). *Economics of the Family and Farming Systems in Sub-Saharan Africa.* Boulder, CO: Westview Press.

Smith, D. and Pearson, R.A. (2005). A review of the factors affecting the survival of donkeys in semi-arid regions of sub-Saharan Africa. *Tropical Animal Health and Production* 37, Supplement 1, 1–9. DOI: 10.1007/s11250-005-9002-5.

Sosovele, H. (1994). Transfer of animal traction technology: Cultural and social issues in Tarime District, Tanzania. In: Starkey, P., Mwenya, E. and Stares, J. (eds.) *Improving Animal Traction Technology: Proceedings of the First Workshop of the Animal Traction Network for Eastern and Southern Africa (ATNESA) Held 18–23*

January 1992, Lusaka, Zambia. Wageningen, Netherlands: Animal Traction Network for Eastern and Southern Africa, 318–20.

Sosovele, H. (2004). Donkey traction in Tanzania: Some critical issues. In: Fielding, D. and Starkey, P. (eds.) *Donkeys, People and Development: A Resource Book of the Animal Traction Network for Eastern and Southern Africa (ATNESA)*. Wageningen, Netherlands: Technical Centre for Agricultural and Rural Cooperation (CTA), 107–12 [web edition pagination].

Starkey, P. (1981). *Farming with Work Oxen in Sierra Leone*. Freetown, Sierra Leone: Sierra Leone Work Oxen Project.

Starkey, P. (1989). *Harnessing and Implements for Animal Traction: An Animal Traction Resource Book for Africa*. Braunschweig, Germany: Vieweg, for GATE/GTZ.

Starkey, P. (1991). Animal traction: Constraints and impact among African households. In: Haswell, M. and Hunt, D. (eds.) *Rural Households in Emerging Societies*. Oxford: Berg, 77–90.

Starkey, P. (1992). Alive and pulling. *Ceres* 24/3, 18–23.

Starkey, P. (1994a). A world-wide view of animal traction highlighting some key issues in eastern and southern Africa. In: Starkey, P., Mwenya, E. and Stares, J. (eds.) *Improving Animal Traction Technology: Proceedings of the First Workshop of the Animal Traction Network for Eastern and Southern Africa (ATNESA) Held 18–23 January 1992, Lusaka, Zambia*. Wageningen, Netherlands: Animal Traction Network for Eastern and Southern Africa, 66–81.

Starkey, P. (1994b). Donkey utilisation in sub-Saharan Africa: Recent changes and apparent needs. In: Bakkoury, M. and Prentis, R.A. (eds.) *Working Equines: Proceedings of Second International Colloquium Held 20–22 April 1994, Rabat, Morocco*. Rabat, Morocco: Actes Editions, Institut Agronomique et Veterinaire Hassan ll, 1–7 [online PDF; original page nos. 289–302].

Starkey, P. (1995). The donkey in South Africa: Myths and misconceptions. In: Starkey, P. (ed.) *Animal Traction in South Africa: Empowering Rural Communities*. Alice, South Africa: South African Network of Animal Traction (SANAT), 139–51.

Starkey, P. (2011). *Livestock for Traction and Transport: World Trends, Key Issues and Policy Implications*. AGA Working Paper Series. Rome: Food and Agriculture Organization of the United Nations, Animal Production and Health Division.

Starkey, P. and Apetofia, K.V. (1986). *Integrated Livestock Systems in Nepal and Indonesia: Implications for Animal Traction Programs in West Africa*. Gainesville, FL: University of Florida, and Washington DC Office of Agriculture and Office of Multisectoral Development.

Steinkeller, P. (2004). The function of written documentation in the administrative praxis of early Babylonia. In: Hudson, M. and Wunsch, C. (eds.) *Creating Economic Order: Record-Keeping, Standardization and the Development of Accounting in the Ancient Near East*. Bethesda, ML: CDL Press, 65–87.

Steinkeller, P. (2007). City and countryside in third millennium southern Babylonia. In: Stone, E.C. (ed.) *Settlement and Society: Essays Dedicated to Robert McCormick Adams*. Los Angeles, CA and Chicago, IL: Cotsen Institute of Archaeology, University of California, Oriental Institute of the University of Chicago, 185–211.

Stol, M. (1995). Old Babylonian cattle. *Bulletin on Sumerian Agriculture* 8, 173–213.

Stone, E. (1999). The constraints on state and urban form in ancient Mesopotamia. In: Hudson, M. and Levine, B.A. (eds.) *Urbanization and Land Ownership in the Ancient Near East*. Cambridge, MA: Peabody Museum of Archaeology and Ethnology, Harvard University, 203–27.

Styring, A.K., Charles, M., Fantone, F., Hald, M.M., McMahon, A., Meadow, R.H., Nicholls, G.K., Patel, A.K., Pitre, M.C., Smith, A., Sołtysiak, A., Stein, G., Weber, J.A., Weiss, H. and Bogaard, A. (2017). Isotope evidence for agricultural extensification reveals how the world's first cities were fed. *Nature Plants* 3/17076, 1–11. DOI: 10.1038/nplants.2017.76.

Sylwander, L. (1994). Women and animal traction technology. In: Starkey, P., Mwenya, E. and Stares, J. (eds.) *Improving Animal Traction Technology: Proceedings of the First Workshop of the Animal Traction Network for Eastern and Southern Africa (ATNESA) Held 18–23 January 1992, Lusaka, Zambia.* Wageningen, Netherlands: Animal Traction Network for Eastern and Southern Africa, 260–5.

Tabbaa, D. (2003). Use of equines and their management in Syria: Historical and modern. In: Pearson, R.A., Fielding, D. and Tabbaa, D. (eds.) *Fourth International Colloquium on Working Equines: Proceedings of a Colloquium Organised by the Society for the Protection of Animals Abroad (SPANA) and the Faculty of Veterinary Medicine, Al Baath University, Hama, Syria, 20–26 April 2002.* Wageningen, Netherlands: Wageningen Academic Publishers, 2–6.

Tani, Y. (1996). Domestic animal as serf: Ideologies of nature in the Mediterranean and the Middle East. In: Ellen, R. and Fukui, K. (eds.) *Redefining Nature: Ecology, Culture and Domestication.* Oxford: Berg, 387–416.

Tarr, J.A. (1999). A note on the horse as an urban power source. *Journal of Urban History* 25, 434–8. DOI: 10.1177/009614429902500306.

Tibbs, P. (1989). *The Importance of Working Animals to the Livestock and Arable Sectors of Modern Agriculture in the People's Republic of China: The Report of a Fellowship Study in the People's Republic of China, Autumn 1987.* Streatley, Berks: Arkleton Trust.

Tiffen, M. (1976). *The Enterprising Peasant: Economic Development in Gombe Emirate, North Eastern State, Nigeria, 1900–1968.* London: HMSO.

Tisserand, J.-L. and Pearson, R.A. (2003). Nutritional requirements, feed intake and digestion in working donkeys: A comparison with other work animals. In: Pearson, R.A., Lhoste, P., Saastamoinen, M. and Martin-Rosset, W. (eds.) *Working Animals in Agriculture and Transport: A Collection of Some Current Research and Development Observations.* EAAP Technical Series No. 6. Wageningen, Netherlands: Wageningen Academic Publishers, 63–74.

Twerda, M., Fielding, D. and Field, C. (1997). Role and management of donkeys in Samburu and Turkana pastoralist societies in northern Kenya. *Tropical Animal Health and Production* 29/1, 48–54. DOI: 10.1007/BF02632348.

Uerpmann, H.-P. (1991). *Equus africanus* in Arabia. In: Meadow, R.H. and Uerpmann, H.-P. (eds.) *Equids in the Ancient World,* Vol. 2. Wiesbaden: Reichert, 12–33.

Ur, J.A. (2010). Cycles of civilization in Northern Mesopotamia, 4400–2000 BC. *Journal of Archaeological Research* 18/4, 387–431. DOI: 10.1007/s10814-010-9041-y.

Vall, E., Dongmo Ngoutsop, A.L., Abakar, O. and Meyer, C. (2002). La traction animale dans le nouveau contexte des savanes cotonnières du Tchad, du Cameroun et de la République centrafricaine. I. Diffusion de la traction animale et sa place dans les exploitations. *Revue d'Elevage et de Médecine Vétérinaire des Pays Tropicaux* 55, 117–28.

van Dijk, L. (2011). *The role of working animals in livelihood security.* Rome: Food and Agriculture Organization of the United Nations. Gateway to Farm Animal Welfare Electronic Consultation: Role, Impact and Welfare of Working (Transport and Traction) Animals, 1–28 February 2011.

Vigne, J.-D. and Helmer, D. (2007). Was milk a "secondary product" in the Old World Neolithisation process? Its role in the domestication of cattle, sheep and goats. *Anthropozoologica* 42/2, 9–40.

Vila, E. (2006). Data on equids from late fourth and third millennium sites in Northern Syria. In: Mashkour, M. (ed.) *Equids in Time and Space: Papers in Honour of Véra Eisenmann*. Oxford: Oxbow, 101–23.

Vila, E. (2014). L'âne domestique en Syrie: réflexions sur les données nouvelles de Mari. In: Butterlin, P., Margueron, J.-C., Muller, B., Al-Maqdissi, M., Beyer, D., Cavigneaux, A. (eds.) *Mari, ni Est, ni Ouest: actes du colloque "Mari, ni Est ni Ouest" tenu les 20–22 octobre 2010 à Damas, Syrie*. Supplément Syria II, Vol. 1. Beyrouth: Presses de l'Ifpo, 425–36.

Vilà, C., Leonard, J.A. and Beja-Pereira, A. (2006). Genetic documentation of horse and donkey domestication. In: Zeder, M.A., Bradley, D.G., Emshwiller, E. and Smith, B.D. (eds.) *Documenting Domestication: New Genetic and Archaeological Paradigms*. Berkeley, CA: University of California Press, 342–53.

Waithanji, E.M. (2009). *Findings of a Preliminary Anthropological Survey on Donkey Welfare in Mwea, Lari, and Limuru, Kenya*. Nairobi: KENDAT.

Waithanji, E.M. (2014). *A brief on gender and working equids' welfare in Kenya*. Paper presented at the launch of The Brooke Voices from Women International Report: Invisible Helpers: Women's Views on the Contributions of Working Donkeys, Horses and Mules to Their Lives, Held 14 May 2014, Portcullis House, London.

Wapnish, P. and Hesse, B. (1991). Faunal remains from Tell Dan: Perspectives on animal production at a village, urban and ritual center. *Archaeozoologica* 4/2, 9–86.

Way, K.C. (2010). Assessing sacred asses: Bronze Age donkey burials in the Near East. *Levant* 42/2, 210–25. DOI: 10.1179/175638010X12797246583852.

Weber, J.A. (1997). Faunal remains from Tell es-Sweyhat and Tell Hajji Ibrahim. In: Zettler, R.L. (ed.) *Subsistence and Settlement in a Marginal Environment: Tell es-Sweyhat, 1989–1995 Preliminary Report*, Vol. 14. Philadelphia, PA: Museum Applied Science Center for Archaeology, University of Pennsylvania Museum of Archaeology and Anthropology, 133–68.

Weber, J.A. (2008). Elite equids: Redefining equid burials of the mid-to late 3rd millennium BC from Umm el-Marra, Syria. In: Vila, E., Gourichon, L., Choyke, A.M. and Buitenhuis, H. (eds.) *Archaeozoology of the Near East VIII: Proceedings of the Eighth International Symposium on the Archaeozoology of Southwestern Asia and Adjacent Areas*. Lyon: Maison de l'Orient et de la Méditerranée, 499–520.

Wilkinson, P.F. (1972). Oomingmak: A model for man-animal relationships in prehistory. *Current Anthropology* 13, 23–44. DOI: 10.1086/201247.

Wilkinson, T.C. (2016). *Steps Toward the Study of Seasonality and Trade*. ArchAtlas, Version 4.1. www.archatlas.org/occpaper/Wilkinson.php [accessed 30 July 2017].

Wilson, R.T. (1978). Studies on the livestock of southern Dafur, Sudan VI. Notes on equines. *Tropical Animal Health and Production* 10, 183–9. DOI: 10.1007/BF02235338.

Wylie, A. (2002). *Thinking from Things*. Berkeley, CA: University of California Press.

Yilmaz, O. (2012). *Domesticated Donkey*. Ankara: Veni Vidi Vici Publications.

Yousef, M.K., Dill, D.B. and Freeland, D.V. (1972). Energetic cost of grade walking in man and burro, *Equus asinus*: Desert and mountain. *Journal of Applied Physiology* 33, 337–40. DOI: 10.1152/jappl.1972.33.3.337.

Zarins, J. (1978). The domesticated Equidae of third millennium BC Mesopotamia. *Journal of Cuneiform Studies* 30, 3–17. DOI: 10.2307/1359930.

Zarins, J. (2014). *The Domestication of Equidae in Third-Millennium BCE Mesopotamia*. Bethesda, MD: CDL Press.

Zeder, M.A. (1986). The equid remains from Tal-e Malyan, southern Iran. In: Meadow, R.H. and Uerpmann, H.-P. (eds.) *Equids in the Ancient World*, Vol. 1. Wiesbaden: Reichert, 366–412.

Zenebe, S. and Fekade, T. (2004). The role of donkey pack-transport in the major grain market of Addis Ababa. In: Fielding, D. and Starkey, P. (eds.) *Donkeys, People and Development: A Resource Book of the Animal Traction Network for Eastern and Southern Africa (ATNESA)*. Wageningen, Netherlands: Technical Centre for Agricultural and Rural Cooperation (CTA), 69–76 [web edition pagination].

Index

For Product Safety Concerns and Information please contact our EU
representative GPSR@taylorandfrancis.com
Taylor & Francis Verlag GmbH, Kaufingerstraße 24, 80331 München, Germany

www.ingramcontent.com/pod-product-compliance
Lightning Source LLC
LaVergne TN
LVHW011132300926
849160LV00008B/616